INTRODUCTORY
ALGEBRA
FOR BEGINNERS

ATEF ZAKI

FOR INFORMATION CONTACT THE AUTHOR AT:
Jupitertut2@aol.com

Library of Congress Control Number: 2010908805

ISBN: 9781453640999

Printed in the United States

First printing 2010

INTRODUCTION

The history of algebra dates back almost 4000 years. It began in ancient Egypt, and later was developed by the Babylonians, Greeks, Hindus, and the Europeans. The root of the word algebra is derived from the Arabic word "Al-jabr."

Algebra is the branch of mathematics that includes the study of rules, operations, reducing equations to the simplest form, and solving the equations to find the value(s) of the unknown(s).

This book was designed to prepare students who have little or no knowledge of algebra to understand the basics and to prepare them for more advanced mathematics. Learning algebra requires time and patience. My advice is to relax, learn the basic concepts, and find time to practice.

This book explains the fundamentals of algebra in a simple way, with easy to understand terms. It includes about 440 examples, as well as the answers to more than 1,330 exercise problems.

DEDICATION

To my family

The most important people in my life

Table of Contents

CHAPTER 3

CHAPTER 4

CHAPTER 5

CHAPTER 8

CHAPTER 9

CHAPTER 12

CHAPTER 1

BASIC SYMBOLS

The following are some of the basic symbols used in algebra:

SYMBOL	MEANING
$+$	Plus (addition)
$-$	Minus (subtraction)
x	Times (multiplication)
$\div$ or $/$	Divided by (Division)
$<$	Less than
$>$	Greater than
$=$	Equals
$\leq$	Less than or equal to
$\geq$	Greater than or equal to
$\sqrt{}$	Square root
$\sqrt[3]{}$	Cubic root
i	Imaginary number
$\pm$	Plus or minus
$\approx$	Approximately equals
$\%$	Per cent
$\not<$	Not less than
$\not>$	Not greater than

SYMBOL	MEANING
$\neq$	Not equal to
y^2	y squared
y^3	y cubed
y^n	y to the power n
$\vert \; \vert$	Absolute value of the number or term between the two vertical lines
$\therefore$	Therefore

Basic Algebraic Operations

The basic algebraic operations include multiplication, division, addition, and subtraction.

The following are the symbols used to indicate that the numbers or letters are multiplied.

Example 1:   10 x 2

Example 2:   9 . 3

Example 3:   7 (6)

Example 4:   (5) 12

Example 5:   (4) (11)

Example 6: x y

Example 7: x y z

Example 8:   (a b) (c d)

Example 9:   (a b c) (x y z)

The answer to a multiplication operation is called "product".
For example the product of 5 (10) is "50".

The following are symbols used to indicate that the numbers or letters are divided:

Example 1:   10 ÷ 2

Example 2:   16 / 4

Example 3: $\frac{18}{9}$

Example 4:   y / x

Example 5:   (a b) ÷ (c d)

Example 6:   (a b c)
              (x y z)

The answer to a division operation is called "**quotient**".
For example the quotient of 10 / 2 is "5".

There is only one symbol used for addition, which is "+", and one symbol used
for subtraction, which is "–".

Example 1:   9 + 5 = 14

Example 2:   14 – 5 = 9

Example 3:   (12 . 5) + (3 . 5)

Example 4:   (9 / 4) – (3 / 4)

Example 5:   x + y

Example 6:   y – x

Example 7:   (a b) + (c d)

Example 8:   (y / z) – (a / b)

The symbol used to indicate that a number or a letter is greater than another
number or a letter is ">"

Example 1:   9 > 8

Example 2:   10 (2) > 7 (2)

Example 3:   (45 / 3) > (15 / 3)

Example 4:   y > x

Example 5:   (a b) > (c d)

Example 6: $(y / x) > (a / z)$

The symbol used to indicate that a number or a letter is less than another number or a letter   is "<"

Example 1: $16 < 17$

Example 2: $6 (3) < 8 (3)$

Example 3: $x < y$

Example 4: $(x \ y) < (y \ z)$

The symbol used to indicate that a number or a letter is not equal to another number or a letter   is "$\neq$"

Example 1: $3 \neq 2$

Example 2: $x \neq y$

Example 3: $(a \ b) \neq (c \ d)$

Example 4: $(y / x) \neq (b / a)$

The symbol used to indicate that a number or a letter is not less than another number or a letter   is "$\not<$".

Example 1: $30 \not< 29$

Example 2: $x \not< y$

Example 3: $(y \ z) \not< (b \ c)$

Example 4: $(c / a) \not< (x / y)$

The symbol used to indicate that a number or a letter is not greater than another number or a letter   is "$\not>$".

Example 1: $5 \not> 10$

Example 2: $x \not> y$
Example 3: $(x \ y) \not> (b \ c)$

Example 4: $(y / z) \not> (c / a)$

The symbol used to approximate an answer is "$\approx$".

Example 1:   4.999 is ≈ 5 (approximately equals 5)

Example 2: $x \approx y$

The **per cent** symbol "%" is used to indicate a relationship between two numbers and it means for every 100. For example if you deposit $100.00 in a savings account, and the **interest** rate is 3% per year, this means that after one year you will receive $100 \ x \ \dfrac{3}{100} = \$3.00$ in interest. If the amount deposited was $10,000.00, after one year you will receive $10{,}000 \ x \ \dfrac{3}{100} = \$300.00$

Example 1:   If the sales tax is 8%, this means that you pay a sales tax of $8.00 for every hundred dollars you spend. If Caroline buys a car for $30,000.00, she will pay a sales tax equal to $30{,}000 \ x \ \dfrac{8}{100} = \$2400.00$

Example 2:   If 10% of the American people are rich. If the total population of the U.S. is 300 million, this means that the number of rich people in the U.S. equals $300{,}000{,}000 \ x \ \dfrac{10}{100}$, which equals 30,000,000 people.

Example 3:   Lydia bought a bicycle for $150.00, and the special offer
said that the buyer will receive a rebate of 20% by mail.
How much will Lydia receive?

$$150 \text{ x } \frac{20}{100} = \$30.00$$

To indicate that a number is raised to a certain **power**, the power is shown as a number close to the top right side of the number.

Example 1: 4^2, this is read four squared, and it means 4 x 4

Example 2: 2^3, this is read two cubed, and it means 2 x 2 x 2.

Example 3: 5^4, this is read five to the power four, and it means
5 x 5 x 5 x 5

Example 4: y^n, this is read y to the power n, and it means y multiplied
by itself n times.

Exercise 1.1

Find the answers to the following operations:

1-	$5 - 3 + 11$		2-	$-6 - 9 - 5$
3-	$7 + (8 - 2)$		4-	$9 + (7 - 12)$
5-	$13 - (4 - 3)$		6-	$-8 - (-2 - 14)$
7-	$3 (8)$		8-	$5 (-8)$
9-	$-5 (9)$		10-	$-6 (-7)$
11-	$3 (-4) 5$		12-	$2 (-9) (-4)$
13-	$-6 (9) (-3)$		14-	$-10 (-2) (-7)$
15-	$(4 \text{ x } 2) - 1$		16-	$16 - (3 \text{ x } 4)$

17-	$18 - 3 (- 6)$	18-	$15 + 3 (- 8 - 3)$
19-	$- 14 + 2 (- 7 - 3)$	20-	$12 - 5 (- 2) + 2 (11)$
21-	$- 24 \div 6$	22-	$- 40 \div (- 4)$
23-	$(8 \text{ x } 9) \div 12$	24-	$(10 \text{ x } 9) \div (- 5)$
25-	$7 (6 - 10) \div (- 14)$	26-	$120 \div (5 \text{ x } 8)$
27-	$36 \div 2 (- 3)$	28-	$(-72) \div 18 (- 2)$
29-	$(13 + 5) \div (- 9)$	30-	$(25 - 4) \div 3$
31-	$(14 - 2) \div (- 3)$	32-	$(7 - 14) \div (8 - 15)$
33-	$10 (5 - 9) - 20 \div (14 + 16)$	34-	$[12 (2 - 7) - 21] \div (4 - 13)$

Exercise 1.2

Write all the correct symbols between the following numbers to reflect how the numbers compare to each other. For example 7, 5 write $7 > 5$ and $7 \not< 5$ and $7 \neq 5$:

1-	10, 8	2-	4, 9
3-	3, 3	4-	$11, (3 \text{ x } 2)$
5-	$13, (2 \text{ x } 7)$	6-	$16, (8 \text{ x } 2)$
7-	$5, (21 \div 7)$	8-	$12, (18 - 6)$
9-	0, 3	10-	$0, - 4$
11-	6, 0	12-	$- 5, 0$
13-	0, 0	14-	$0, (3 - 10)$
15-	$14, - (4 / 2)$	16-	$-15, (12 / 3)$
17-	$-7, (- 10 / 2)$	18-	$12, (3 \text{ x } 4)$
19-	$-9, (-10 - 3)$	20-	$15, (10 / 2)$

Exercise 1.3

Write the following letters and numbers using the correct symbols to reflect the indicated criteria:

1-	y plus x	2-	z minus a
3-	a times b	4-	c divided by y
5-	y is less than b	6-	b is greater than c
7-	d is equal to x	8-	y is less than or equal to a

9- b is greater than or equal to z

10-	Square root of 16	11-	cubic root of 8
12-	x plus or minus (y + z)	13-	twenty five percent

14- 255 is not less than 245

15- (x + y) is not greater than (a + b)

16- (b + c) / (y + z) is not equal to (a + b) / (x + y)

17-	z squared	18-	b cubed
19-	10 to the power 4	20-	The absolute value of 6

21- The absolute value of minus fifteen

22- The absolute value of (y / d)

CHAPTER 2

REAL NUMBERS

Positive and negative numbers such as 0, 1, 2, 3, … and –1, –2, –3, … are called **whole numbers**.

Whole numbers can be represented graphically by drawing a straight line. Choose a point on the line (which is called an axis) and call it 0. This point is referred to as the **origin**. Locate another point on the line at a certain distance on the right side of the origin, and call this point 1. Using the same distance between the point of origin and point 1, locate another point on the line on the right side of point 1, and call this point 2. Follow this same procedure to locate all other required numbers on the line. See figure 2.1.

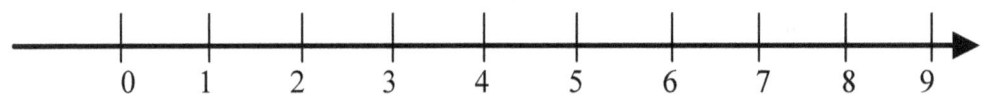

FIGURE 2.1

The whole numbers shown in figure 2.1 above are all positive numbers. Negative numbers such as –1, –2, –3 etc. can also be shown on the line to the left of the point of origin. See figure 2.2.

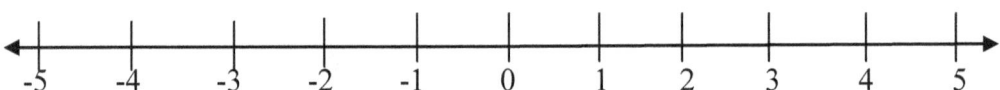

FIGURE 2.2

The horizontal line shown in figure 2.2 is called the x-**axis**. Similarly, the positive and negative numbers can also be shown on a vertical line and it is called the **y-axis**. See figure 2.3

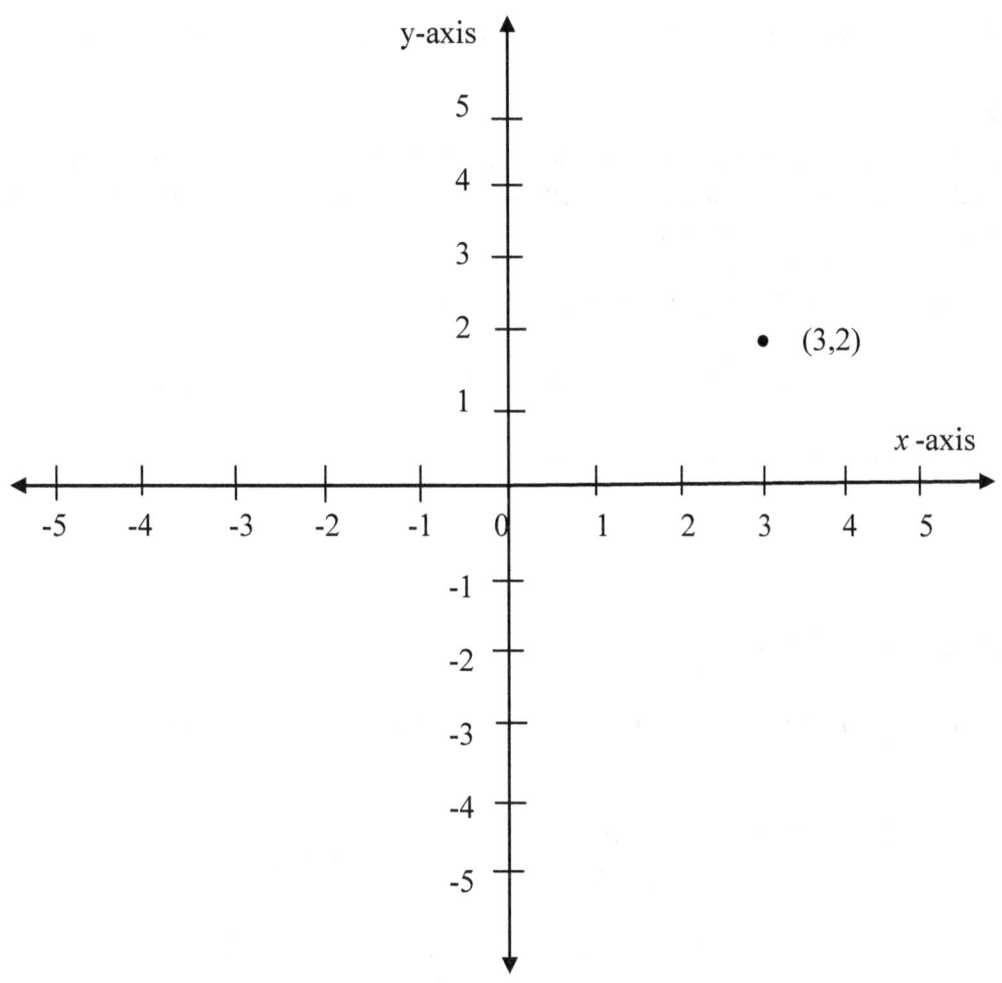

FIGURE 2.3

The x - **coordinate** of a point is the value that determines how far the point is from the origin on the x -axis. The **y- coordinate** of a point is the value that determines how far the point is from the origin on the y-axis. For example the point shown in figure 2.3 above has an x - coordinate of 3 and a y- coordinate of 2.

The set of positive and negative whole numbers is called the **set of integers**

Rational Numbers

A rational number is a number that is the quotient of dividing one integer by another, no matter whether the integer is positive or negative, with the exception of having 0 in the denominator.

Example 1:   1 / 5 = 0.2 (0.2 is a rational number)

Example 2:   4 / 1 = 4 (4 is a rational number)

Example 3:   3 / 2 = 1.5 (1.5 is a rational number)

Example 4:   –21 / 3 = –7 (–7 is a rational number)

Irrational Numbers

An irrational number is a number that does not have an exact definite value.

Example 1:   The square root of 2
$\sqrt{2}$ = 1.414213562

The answer has an indefinite number of decimals; therefore the answer is an irrational number.

Example 2:   The Greek letter π is used in mathematics to indicate the value of the length of the circumference of a circle divided by the diameter of the circle.
$\pi = 3.14159$

The value of π has an indefinite number of decimals; therefore π is an irrational number.

Real Numbers

Real numbers can be rational numbers, or irrational numbers, and can be positive, negative or zero.

Examples: 0
 9
 24
 -15
 a b
 y / z (z is not equal to zero)
 $-xy$

Absolute Value

The symbol used for absolute value of a number or an unknown is two vertical lines with the number or the unknown between them.

The absolute value of an unknown such as x, written $|x|$,
is equal to x, if x is ≥ 0
or is equal to $-x$, if x is < 0

Example 1: $|6| = 6$

Example 2: $|-18| = -(-18) = 18$

Example 3: $|9-13| = |-4| = -(-4) = 4$

Example 4: $|9/3| = |3| = 3$

Example 5: $|-35/7| = |-5| = -(-5) = 5$

Example 6: $|2x/2| = |x| = x$

Example 7: $|-4y \cdot 3| = |-12y| = -(-12y) = 12y$

Additive Inverse

An additive inverse is a number when added to another number makes the sum of the two numbers equals zero.

Example 1: 7 is the additive inverse of -7 since $(7 - 7 = 0)$

Example 2: -9 is the additive inverse of 9 since $(-9 + 9 = 0)$

Example 3: $-b$ is the additive inverse of b since $(-b + b = 0)$

Example 4: $5xy$ is the additive inverse of $-5xy$
since $(5xy - 5xy = 0)$

Properties of Whole Numbers:

Addition of Whole Numbers

Commutative Law of Additions

The sum of any two real numbers will always be the same regardless of the order in which the numbers are added.

Example 1: $x + y = y + x$

Example 2: $3 + 4 = 4 + 3 = 7$

Example 3: $-2 + 5 = 5 - 2 = 3$

Example 4: $(a + cy) = (cy + a)$

Example 5: $(xyz + abc) = (abc + xyz)$

Associative Law of Additions

The sum of any three real numbers will always be the same regardless of how the numbers are grouped.

Example 1: $x + (y + z) = (x + y) + z$

Example 2: $6 + (7 + 8) = (6 + 7) + 8 = 21$

Example 3: $9 + (-3 + 1) = 9 + 1 - 3 = 7$

Example 4: $(ab + cd + yz) = cd + (ab + yz)$

Example 5: $(xyz + aby + czy) = czy + (aby + xyz)$

The Additive Identity

If zero is added to any number the answer is the number.

Example 1: $0 + x = x + 0 = x$

Example 2: $y + 0 = 0 + y = y$

Example 3: $0 + 12 = 12 + 0 = 12$

Example 4: $11 + 0 = 0 + 11 = 11$

Example 5: $0 + ab = ab + 0 = ab$

Simplifying fractions

Fractions can be simplified when the numerator and denominator have common factors. The numerator and denominator can be divided by the common factors and the resulting fraction is exactly equal to the original fraction and is called the **equivalent fraction**.

Example 1:   Simplify the fraction $\dfrac{xy}{xz}$

Note that the numerator and denominator have a common factor (x).

Divide the numerator and denominator by the common factor (x).

The resulting equivalent fraction is $\dfrac{y}{z}$

Example 2:   Simplify the fraction $\dfrac{12x^2 yz^2}{2xyz}$

Note that the numerator and denominator have several common factors including 2, x, y, and z.

Divide the numerator and denominator by 2

The resulting fraction will be $\dfrac{6x^2 yz^2}{xyz}$

Divide the numerator and denominator by (x y z)

The **reduced equivalent fraction** is $6\,x\,z$

Addition and subtraction of fractions

Fractions that have the same denominator can be added or subtracted
By writing the denominator (the lower number of a fraction) they have in common and adding or subtracting the numerators (the upper numbers of a fraction).

Example 1: $2/9 + 5/9 = (2+5)/9 = 7/9$

Example 2: $6\,a/b - 4\,a/b = (6\,a - 4\,a)/b = (2\,a)/b$

Example 3: $[y / (a + b)] + [z / (a + b)] = (y + z) / (a + b)$

Example 4: $[(2x - y) / (y - z)] - [(x + y) / (y - z)]$

$$= [(2x - y) - (x + y)] / (y - z)$$

$$= (2x - x - y - y) / (y - z)$$

$$= (x - 2y) / (y - z)$$

If the fractions do not have the same denominator, addition and subtraction can be done by finding the least common denominator of all the denominators.

The **least common denominator** is defined as the least common number of the numbers in the denominators that is divisible by each number in all the denominators.

To find the least common denominator:

1- Factor each denominator to its prime numbers, for example
$(1 / 6) + (2 / 8) + (3 / 48)$
The denominator 6 is factored to 2 . 3
The denominator 8 is factored to 2 . 2 . 2
The denominator 48 is factored to 2 . 2 . 2 . 2 . 3

2- Count the largest number of times each prime appears in the denominators.

The largest number of times the number 2 appears is 4
The largest number of times the number 3 appears is 1

3- The least common denominator is the product of all the largest number of times each prime number appears. This means 2 . 2 . 2 . 2 . 3 = 48

Therefore, the least common denominator of the three denominators (6, 8, and 48) is 48.

To add or subtract the fractions, transform the fractions into new fractions using the least common denominator for each one of them. To do that divide the least common denominator by the denominator of the first fraction, then multiply the quotient by the numerator. Repeat the process for all other fractions.

48 / 6 = 8, multiply 8 by the numerator (1), therefore, the first fraction becomes 8 / 48. Repeating the process, the second fraction becomes 12 / 48, and the third fraction becomes 3 / 48.

To add or subtract the fractions, write a fraction whose denominator is the least common denominator, and the numerator is the new numerators of all the fractions with their respective signs.

Therefore, (1 / 6) + (2 / 8) + (3 / 48) = (8 + 12 + 3) / 48 = 23 / 48

Example 1:   Add the following fractions: (3 / 5) + (5 / 15) + (2 / 25)

Factor the denominators
5 = 1 . 5
15 = 3 . 5
25 = 5 . 5

The least common denominator = 3 . 5 . 5 = 75

(75 ÷ 5) 3 = 45, therefore the first fraction becomes 45 / 75

(75 ÷ 15) 5 = 25, therefore the second fraction becomes 25 / 75

(75 ÷ 25) 2 = 6, therefore the third fraction becomes 6 / 75

(45 / 75) + (25 / 75) + (6 / 75) = (45 + 25 + 6) / 75 = 76 / 75

Example 2:   Add the following fractions: (2 / 6) + (5 / 12) + (7 / 18)

Factor the denominators
6 = 2 . 3
12 = 2 . 2 . 3
18 = 2 . 3 . 3

The least common denominator = 2 . 2 . 3 . 3 = 36

(36 ÷ 6) 2 = 12, therefore the first fraction becomes 12 / 36

(36 ÷ 12) 5 = 15, therefore the second fraction becomes 15 / 36

(36 ÷ 18) 7 = 14, therefore the third fraction becomes 14 / 36

(2 / 6) + (5 / 12) + (7 / 18) = (12 / 36) + (15 / 36) + (14 / 36)
= 41 / 36

Example 3: Subtract the following fractions: (9 / 10) – (7 / 12) – (1 / 6)

Factor the denominators
10 = 2 . 5
12 = 2 . 2 . 3
6 = 2 . 3

The least common denominator = 2 . 2 . 3 . 5 = 60
(60 ÷ 10) 9 = 54, therefore the first fraction becomes 54 / 60

(60 ÷ 12) (– 7) = – 35, therefore the second fraction becomes
– 35 / 60

(60 ÷ 6) (– 1) = 10, therefore the third fraction becomes – 10 / 60

(54 / 60) – (35 / 60) – (10 / 60) = (54 – 35 – 10) / 60 = 9 / 60
= 3 / 20

Example 4: Subtract the following fractions: (8 / 5) – (6 / 15) – (6 / 30)

Factor the denominators
5 = 1 . 5
15 = 3 . 5
30 = 2 . 3 . 5

19

The least common denominator = 2 . 3 . 5 = 30

(30 ÷ 5) 8 = 48, therefore the first fraction becomes 48 / 30

(30 ÷ 15) (– 6) = – 12, therefore the second fraction becomes – 12 / 30

(30 ÷ 30) (– 6) = – 6, therefore the third fraction becomes – 6 / 30

(8 / 5) – (6 / 15) – (6 / 30) = (48 / 30) – (12 / 30) – (6 / 30)
= (48 – 12 – 6) / 30 = 30 / 30 = 1

Multiplication of Whole Numbers

Commutative Law of Multiplications

The product of any two real numbers will always be the same regardless of the order in which the numbers are multiplied.

Example 1: $x \, y = y \, x$

Example 2:   3 x 4 = 4 x 3 = 12

Example 3:   – 2 x 5 = 5 x – 2 = – 10

Example 4:   – b . a = a . – b

Associative Law of Multiplication

The product of any three real numbers will always be the same regardless of how the numbers are grouped.

Example 1: $x \, (y \, z) = (x \, y) \, z = (x \, z) \, y = x \, y \, z$

Example 2:   4 (3 x 2) = (4 x 3) 2 = (4 x 2) 3 = 24

Example 3:   9 (− 3 x 1) = 9 x 1 x − 3 = − 27

Example 4:   a b c = a c b = b a c = b c a = c a b = c b a

The Distributive Law of multiplication

If similar terms are added to each other, such as 4d and 8 d, the sum of these numbers can be simplified by factoring the similar numbers.

Example 1: $4 d + 8 d = d (4 + 8) = 12 d$

Example 2: $3 y + 5 y + 7 y = y (3 + 5 + 7) = 15 y$

Example 3: $2 a b + 6 a b + 12 a b = a b (2 + 6 + 12) = 20 a b$

Example 4: $2 x y z + 7 x y z + 10 x y z = x y z (2 + 7 + 10) = 19 x y z$

The same rule applies to terms that are subtracted from each other.

Example 1: $12 y − 3 y = y (12 − 3) = 9 y$

Example 2: $9 x − 3 x − 2 x = x (9 − 3 − 2) = 4 x$

Example 3: $11 y z − 2 y z − y z = y z (11–2–1) = 8 y z$

Example 4: $− 5 a b c − 6 a b c − 9 a b c = a b c (− 5 − 6 − 9) = − 20 a b c$

The Multiplication Identity

If a number is multiplied by one the answer is the number.

Example 1:   y x 1 = 1 x y = y

Example 2:   5 x 1 = 1 x 5 = 5

Example 3:   − 6 x 1 = 1 x − 6 = − 6

Example 4:   − b x 1 = 1 x − b = − b

Multiplication of Rational Numbers

To multiply rational numbers, multiply the numerators of all rational numbers, and also multiply the denominators of all the rational numbers, then simplify the answer if it can be simplified.

Example 1: $(3 / 4) \cdot (2 / 5) = (3 \cdot 2) / (4 \cdot 5) = 6 / 20 = 3 / 10$

Example 2: $(4 / 7) \cdot (3 / 4) \cdot (5 / 6) = (4 \cdot 3 \cdot 5) / (7 \cdot 4 \cdot 6)$
$= 60 / 168 = 30 / 84 = 15 / 42 = 5 / 14$

Example 3: $(a / b) \cdot (b / c) \cdot (y / z) = (a \cdot b \cdot y) / (b \cdot c \cdot z)$

$= (a \, y) / (c \, z)$

Example 4: $(x y / a) \cdot (3 \, a \, y) / 2 \, b \, z) \cdot (2 \, z / 3 \, c \, y)$

$= (x y \cdot 3 \, a \, y \cdot 2 \, z) / (a \cdot 2 \, b \, z \cdot 3 \, c \, y)$

$= (6 x y \cdot a \, y \cdot z) / (6 \, a \cdot b \, z \cdot c \, y)$

$= x y / b \, c$

Division of Rational Numbers

The **multiplicative inverse (reciprocal)** of a number is defined as 1 divided by the number.

For example, the multiplicative inverse (reciprocal) of 5 = 1 / 5, and the multiplicative inverse (reciprocal) of 1 / 4 = (1) ÷ (1 / 4) = 4

To divide two rational numbers such as (1 / 2) ÷ (1 / 6), keep the first number as it is (1 / 2), then convert the convert the divided by sign (÷) to multiply by sign (x), and write the multiplicative inverse (reciprocal) of the second number (6). The answer will be the first number multiplied by the multiplicative inverse (reciprocal) of the second number, (1 / 2) . (6) = 3

Example 1: Divide (5 / 6) by (2 / 3)
(5 / 6) ÷ (2 / 3) = (5 / 6) . (3 / 2) = 15 / 12 = 5 / 4

Example 2: Divide (3 / 25) by (2 / 100)
(3 / 25) ÷ (2 / 100) = (3 / 25) . (100 / 2) = 300 / 50
= 150 / 25 = 30 / 5 = 6

Example 3: Divide (3 y / 2 z) by (y / 6 z)
(3 y / 2 z) ÷ (y / 6 z) = (3 y / 2 z) . (6 z / y)
= 3 . 3 = 9

Example 4: Divide (2 a b / 6 c d) by (6 b / 3 d)
(2 a b / 6 c d) ÷ (6 b / 3 d) = 2 a b / 6 c d) . (3 d / 6 b)
= (2 a b) (3 d) ÷ (6 c d) (6 b) = 6 a b d ÷ 36 c b d
= a / 6 c

Multiplication and division by zero

When a number, any number, is multiplied by zero the product is always equal to zero.

Examples: 4 x 0 = 0   15 x 0 = 0   − 11 x 0 = 0
             0 x 9 = 0   0 x 0 = 0   0 x − 7 = 0

When a number is divided by zero, the quotient is an undefined number sometimes called (infinity). Dividing zero by zero also results in an undefined quotient. However, the quotient of dividing zero by a number, other than zero, is always equal to zero.

Examples: $\dfrac{10}{0}$ = an undefined number

$\dfrac{0}{0}$ = an undefined number

$\dfrac{0}{3} = 0$

Exercise 2.1

1-   Rewrite (y + z) using the Commutative Law of Additions.
2-   Rewrite 6 + (– 7 + 8) using the Associative Law of Additions, and find the answer.
3-   Rewrite 2 a + 3 b + 4 c using the Associative Law of Additions.
4-   Rewrite y . z using the Commutative Law of Multiplications.
5-   Rewrite   9(8) using the Commutative Law of Multiplications, and find the answer.
6-   Rewrite 8(– 4) (2) using the Associative Law of Multiplications, and find the answer.
7-   Rewrite (3 y . 5 b . 7 c) using the Associative Law of Multiplications
8-   Using the Distributive law of Multiplication find the sum of:
      5 y z + 9 y z + 16 y z
9-   Using the Distributive law of Multiplication find the sum of:
      2 a b c + 3 a b c + 5 a b c
10-   Using the Distributive law of Multiplication find the sum of:
      25 a – 12 a – 6 a
11-   Using the Distributive law of Multiplication find the sum of:
      13 x y – x y – 4 x y
12-   Using the Distributive law of Multiplication find the sum of:
      –17 b y – 7 b y – 6 b y
13-   According to the Multiplication Identity, what should z be multiplied by to get an answer of z.
14-   What is the product of 0 x y
15-   What is the sum of 0 and a?
16-   What is the additive inverse of – 9?
17-   what is the additive inverse of 100?

18-   What is the additive inverse of xyz?
19-   What is the absolute value of -99
20-   What is the absolute value of $(-2z/b)$

Exercise 2.2

Add or subtract the following fractions and simplify:

1-	$(5/4)+(7/8)$	2-	$(7/10)+(9/7)$
3-	$(10/4)-(2/5)$	4-	$(11/12)-(3/4)$
5-	$(2/3)+(3/9)$	6-	$(13/3)-(25/6)$
7-	$(4/7)+(3/5)$	8-	$(16/3)-(3/9)$
9-	$(11/3)-(1/6)-(1/2)$	10-	$(5/6)+(3/8)+(1/10)$
11-	$(3/10+(4/5)+(8/20)$	12-	$(11/12)-(3/16)-(9/28)$
13-	$(1/4)+(1/10)+(1/6)$	14-	$(9/10)+(7/50)-(1/20)$
15-	$(7/3)+(5/6)+(5/12)$	16-	$(15/7)+(5/9)+(2/3)$
17-	$(9/2)-(7/4)-(3/8)$	18-	$(8/3)-(1/6)-(4/9)$
19-	$(13/5)-(3/10)+(4/15)$	20-	$(11/7)+(3/14)-(6/21)$

Exercise 2.3

Multiply the following fractions and simplify:

1-	$(2/3).(3/6)$	2-	$(5/7).(2/10)$
3-	$(3/7).(21/9)$	4-	$(11/12).(18/22)$
5-	$(8/15).(5/4)$	6-	$(4/19).(76/8)$
7-	$(5/11).(22/10)$	8-	$(3/5).(10/6)$
9-	$(1/4).(8/5).(3/2)$	10-	$(14/3).(6/5).(10/7)$
11-	$(13/14).(7/26).(20/3)$	12-	$(2/3).(1/6).(4/9)$
13-	$(7/5).(25/35).(5/4)$	14-	$(11/10).(9/8).(7/22)$
15-	$(2/3).(17/8).(12/34)$	16-	$(19/5).(2/38).(6/10)$
17-	$(1/2).(7/6).(2/3)$	18-	$(9/4).(3/5).(7/20)$
19-	$(4/7).(5/11).(22/10)$	20-	$(2/3).(5/6).(9/6)$

Exercise 2.4

Divide the following fractions and simplify:

1- $(3 / 4) \div (3 / 2)$

2- $(9 / 5) \div (3 / 5)$

3- $(8 / 9) \div (2 / 3)$

4- $(14 / 15) \div (7 / 2)$

5- $(4 / 7) \div (21 / 5)$

6- $(16 / 11) \div (8 / 33)$

7- $(12 / 26) \div (6 / 13)$

8- $(17 / 19) \div (9 / 38)$

9- $(22 / 7) \div (11 / 35)$

10- $(29 / 25) \div (6 / 100)$

11- $(1 / 2)(1 / 5) \div (7 / 20)$

12- $(4 / 3)(3 / 7) \div (6 / 21)$

13- $(1 / 3)(2 / 9) \div (8 / 108)$

14- $(3 / 5)(2 / 7) \div (3 / 35)$

15- $(2 / 3)(11 / 12) \div (2 / 9)$

16- $(11 / 14)(7 / 5) \div (22 / 10)$

17- $(5 / 6)(2 / 3) \div (5 / 18)$

18- $(13 / 15)(3 / 2) \div (2 / 100)$

19- $(4 / 7)(6 / 5) \div (12 / 70)$

20- $(8 / 9)(7 / 4) \div (7 / 45)$

CHAPTER 3

EXPONENTS

An exponent is defined as a quantity or a number that represents the power to which another quantity or number is raised. For example 5^4 (read five to the power 4) means that the number 5 is multiplied by itself four times $(5 \cdot 5 \cdot 5 \cdot 5)$. The number 5 is called **the base**, and the number 4 is called **the exponent**. The base and exponents can be positive quantities or numbers, negative quantities or numbers, or fractions.

When a term does not have an exponent, such as (b), this means that the term, in this case b, is raised to the power 1.

Example 1: 2^6 (2 to the power 6)

Example 2: y^3 (y cubed)

Example 3: x^y (x to the power y)

Example 4: a^{-1} (a to the power -1)

Example 5: $b^{2/3}$ (b to the power two third)

Example 6: z (z to the power 1)

Laws of Exponents:

Zero exponent

If a real number (with the exception of zero) is raised to the power 0, the answer is always 1.

Example 1: $2^0 = 1$

Example 2: $15^0 = 1$

Example 3: $y^0 = 1$

Example 4: $(x + y)^0 = 1$

Example 5: $-5^0 = 1$

Negative Exponent

If a real number (with the exception of zero) is raised to a negative integer number, moving the real number and its negative exponent to the denominator of a fraction that has 1 as a numerator, changes the sign of the exponent from negative to positive.

Example 1: $x^{-1} = 1 / x$

Example 2: $y^{-4} = 1 / y^4$

Example 3: $(x + y)^{-2} = 1 / (x + y)^2$

Example 4: $z^{-3} . z^{-4} = 1 / z^3 . 1 / z^4 = 1 / z^7$

Exponents Product Rule

If two bases are identical but raised to two identical or different exponents, the product of these two terms is the base raised to a power equals the sum of the two exponents.

$$x^b . x^c = x^{(b+c)}$$

Example 1: $4^3 . 4^2 = 4^5$

Example 2: $(-5)^4 (-5)^3 = (-5)^7$

Example 3: $(a + b)^5 (a + b)^4 = (a + b)^9$

Example 4: $(2^2 y^3)(2^4 y^5) = 2^6 y^8$

If a base is raised to an exponent, then the base and the exponent are both raised to a new exponent, to simplify this term, the base remains the same, and the two exponents are multiplied by each other.

$$(y^m)^d = y^{md}$$

Example 1: $(3^2)^5 = 3^{10}$

Example 2: $(a^3)^4 = a^{12}$

Example 3: $[(x+y)^2]^3 = (x+y)^6$

Example 4: $y^{1/4} \cdot y^{5/4} = y^{6/4} = y^{3/2}$

Example 5: $(5^2)^{3/2} = 5^{2.3/2} = 5^3$

If a base consists of terms or numbers multiplied by each other, and the base is raised to one power, the term can be simplified by raising each one of the two terms or numbers to the power.

$$(x\ y)^z = x^z\ y^z$$

Example 1: $(2 \cdot 4)^5 = 2^5 \cdot 4^5$

Example 2: $(7 \cdot 9 \cdot 11)^4 = 7^4 \cdot 9^4 \cdot 11^4$

Example 3: $(x \cdot y \cdot z)^2 = x^2 y^2 z^2$

Example 4: $[(a+b) \cdot c]^3 = (a+b)^3 \cdot c^3$

Example 5: $(3\ y)^{4/3} = 3^{4/3}\ y^{4/3}$

Exponents Quotient Rule

If a base consists of a term or a number (other than zero) raised to a power, and is divided by the same base raised to a different power, the fraction can be simplified by writing the base raised to a power equals the numerator power minus the denominator power.

Exponent quotient rule ....$a^y / a^x = a^{y-x}$

Example 1: $6^8 / 6^3 = 6^{8-3} = 6^5$

Example 2: $19^{4/5} / 19^{1/5} = 19^{4/5-1/5} = 19^{3/5}$

Example 3: $y^7 / y^5 = y^{7-5} = y^2$

Example 4: $(x+2)^{10} / (x+2)^7 = (x+2)^{10-7} = (x+2)^3$

Example 5: $(a+b)^{5/8} / (a+b)^{1/8} = (a+b)^{5/8-1/8} = (a+b)^{4/8} = (a+b)^{1/2}$

If a base consists of terms or numbers divided by each other and the base is raised to a power, the term can be simplified by raising each one of the base terms or numbers to the power.

$$(x/y)^z = x^z / y^z$$

Example 1: $(3/4)^5 = 3^5 / 4^5$

Example 2: $(11/6)^4 = 11^4 / 6^4$

Example 3: $(7/5)^{2/3} = 7^{2/3} / 5^{2/3}$

Example 4: $[(a+b)/c]^3 = (a+b)^3 / c^3$

Example 5: $(y/x)^{1/2} = y^{1/2} / x^{1/2}$

Scientific Notation

Scientific notation, is a method used to express very large or very small numbers in a compact form that is easy to use in computations. In scientific notation, any number is expressed as a number between 1 and 10 multiplied by a power of 10.

Numbers greater than 10 are expressed by positive powers of 10 and numbers less than 1 are expressed by negative powers of 10.

Example 1:   1, 000 can be expressed as 1×10^{3}

Example 2:   10, 000 can be expressed as 1×10^{4}

Example 3:   250,000 can be expressed as 2.5×10^{5}

Example 4:   999,000 can be expressed as 9.99×10^{5}

Example 5:   1, 000, 000 can be expressed as 1×10^{6}

Example 6:   0.02 can be expressed as 2×10^{-2}

Example 7:   0.005 can be expressed as 5×10^{-3}

Example 8:   0.0000064 can be expressed as 6.4×10^{-6}

Exercise 3.1

Simplify the following terms:

1- $2^{0} . 9^{0}$

2- $a^{0} . a^{3}$

3- $5^{-2} . 5^{5}$

4- $4^{1/2} . 4^{5/2}$

5- $7^{3} . 7^{2}$

6- $8 . 4^{2}$

7- $y^{1/10} . y^{4/10}$

8- $x^{1/4} . x^{5/4}$

9- $y^{3} . y^{5}$

10- $b^{5/2} . b$

11- $x^{3} . x^{-1}$

12- $y^{2} (-y^{5})$

13- $-5 b^{3} . 7 b^{4}$

14- $-2 a^{4} b^{3} . 6 a^{3} b$

15- $-3 x^{4} y^{2} . -5 x^{3} y^{3}$

16- $(x-5)^{1/3} . (x-5)^{8/3}$

17- $(2x^2)^3$ 18- $(3x^3 \cdot y^3)^2$

19- $[(x-y)^2]^{3/2}$ 20- $2y^2z^3 \cdot 2y^3z^2$

21- $(b-7)^9(b-7)^4$ 22- $(2a-b)^2 \cdot (2a-b)^4$

23- $[(x+y)^5]^4$ 24- $y^6 \cdot (y^3)^4$

25- $(2y^2 \cdot z^2)(y^{1/2} \cdot z)$ 26- $y^{1/6}(y^{5/6}z^{1/3})$

Exercise 3.2

Simplify the following terms

1- $5 \div 5^{1/2}$ 2- $3^5 \div 3^2$

3- $9^2 \div 9^{1/4}$ 4- $7^{3/2} \div 7^{1/2}$

5- $y^{3/4} \div y$ 6- $x^{5/2} \div x^{3/2}$

7- $12x^3y^4 \div 3x^2y^3$ 8- $6y^5z^4 \div 18y^3z^2$

9- $10a^5b^7 \div 2a^3b^6$ 10- $6x^{4/3}y^{2/3} \div 3x^{1/3}y^{1/3}$

11- $12x^{3/5}y^{4/5} \div 4x^{2/5}y^{3/5}$ 12- $15a^{7/3}b^{5/3} \div 5a^{4/3}b^{2/3}$

13- $8a^4b^6c^3 \div 4ab^4c$ 14- $(3y^{6/5}+y) \div y^{4/5}$

15- $(6x^{7/3}-4x) \div 4x^{1/3}$ 16- $(3x^{1/2}y^{1/3})^6 \div (xy)$

17- $(4/3)^2$ 18- $(9/7)^{2/3}$

19- $(3 \cdot 5)/(3 \cdot 5)^{1/3}$ 20- $2^5 \cdot 3^3 \div 2^2 \cdot 3^2$

21- $4^{5/3}-4^{2/3}/4^{2/3}$ 22- $(2a/3b)^3$

23- $(3yz/4ab)^2$ 24- $(10x^{1/4}y^{3/4})^8 \div (5xy^2)$

25- $[(x-y)/(x+y)]^4$ 26- $[(3ab/5yz)]^2$

Exercise 3.3

Express the following numbers in scientific notation:

1- 64 2- 144

3- 300 4- 550

5- 625 6- 999

7- 1000 8- 1950

9- 7,520 10- 12690

11-	22, 500	12-	167,900
13-	235,500	14-	500,000
15-	5,200,000	16-	7,359,000
17-	12,250, 000	18-	125,600,000
19-	0.752	20-	0.0967
21-	0.0067	22-	0.00086
23-	0.000079	24-	0.0000024
25-	1.0	26-	2.50
27-	6.25	28-	9.749
29-	13.45	30-	17.95

Exercise 3.4

Write the following numbers without scientific notation, and simplify:

1-	0×10^2	2-	1×10^2
3-	1.5×10^2	4-	2.75×10^3
5-	3.24×10^3	6-	12.6×10^3
7-	5.679×10^4	8-	75.23×10^4
9-	119×10^4	10-	275×10^4
11-	0.74×10^5	12-	0.956×10^5
13-	1.4×10^5	14-	5.367×10^5
15-	0.454×10^6	16-	1.5×10^6
17-	6.5432×10^6	18-	25.1234×10^6
19-	6.1×10^{-1}	20-	9.85×10^{-2}
21-	2.2×10^{-3}	22-	6.78×10^{-4}
23-	3.456×10^{-5}	24-	1.98×10^{-6}
25-	$(6 \times 10^3) \div (2 \times 10^{-2})$	26-	$(9 \times 10^{-3}) \div (3 \times 10^2)$
27-	$(8 \times 10^{-4}) \div (2 \times 10^{-2})$	28-	$(5 \times 10^2) \div (3 \times 10^5)$
29-	$4 \div (2 \times 10^{-3})$	30-	$(6 \times 10^{-3}) \div (3 \times 10^3)$

CHAPTER 4

POLYNOMIALS

A polynomial is defined as an expression of **terms** that may include variables and constants. A polynomial must have a definite value. For example, $2x^2 + 4x - 2$ is a polynomial, however, the expression $2x^2 + 3/x - 7x^{2/3}$ is not a polynomial because it includes a term that is divided by a variable x (the second term), and also because it includes the exponent 2/3 (the third term) which is not a whole number.

Polynomials have different names depending on the number of terms it includes:

A polynomial with only one term, such as $6x$ or $12yz$, is called a **monomial**.

A polynomial with two terms, such as $(ab - 3)$, is called a **binomial**.

A polynomial with three terms, such as $(2y^2 - 3y + 2)$, is called a **trinomial**.

A polynomial with more than three terms, such as $(2y^2 x^4 - 6yx^2 + 5x - 9y)$, is called a **multinomial**.

Degree of a Polynomial

Polynomials are written with the terms arranged in a decreasing order, this means the term with the highest exponent comes first, then comes the term with the next highest exponent, and so on.

Example: $3y^4 + 2y^3 - 6y - 5$

To determine the degree of a polynomial look at the highest exponent on the terms. This polynomial has four terms, the first term has an exponent of 4 (fourth degree); the second term has an exponent of 3 (third degree); the third term has an exponent of 1 (first degree), and the fourth term is a constant number.

This means that the polynomial in this example is a fourth degree multinomial since the highest exponent is 4.

If the polynomial terms include more than one unknown raised to different powers such as ($4 x^3 y^2 - 3 x^2 y + 2 x \, y + 9$), this would be called a third degree multinomial in x, and a second degree multinomial in y.

Coefficient

A coefficient is the constant number in a term. For example, the first term of the polynomial $3 y^4 + 2 y^3 - 6 y - 5$ which is ($3 y^4$) includes a coefficient of 3. The coefficient of the second term ($2 y^3$) is 2, and the coefficient of the third term ($- 6 y$) is $- 6$. If there is no coefficient multiplied by a term such as ($x^2 y^2$), this means that the coefficient is 1.

Addition of Polynomials

To add polynomials, add the like terms. **Like terms** are the terms that have the same factors. For example, the terms 2 a b c, 5 c a b, 9 b a c, and 7 a c b are like terms because they all include (a, b, c).

Example 1: Add $6 x - 2$ y, and 7 y $- 3 x$

Combine like terms

$(6 x - 3 x) + (7 y - 2y)$

$= 3 x + 5$ y

Example 2: Add $2 x - 3$ y $+ 5$ z, and $2 x + 5$ y $- 4$ z

Combine like terms

$(2 x + 2 x) + (5 y - 3 y) + (5 z - 4 z)$

$= 4 x + 2$ y $+ z$

Example 3:   Add $5x^2 + 3x$, and $7x^2 + 7x$

Combine like terms

$(5x^2 + 7x^2) + (3x + 7x)$

$= 12x^2 + 10x$

Example 4:   Add $2x^3 + 4x^2 - x$, and $3x^3 - 6x^2 + 2x$

Combine like terms

$(2x^3 + 3x^3) + (4x^2 - 6x^2) + (-x + 2x)$

$= 5x^3 - 2x^2 + x$

Subtraction of Polynomials

To subtract polynomials, use the additive inverse of every term in the polynomial, to be subtracted, then add the like terms. For example to subtract 2 z from 10 z, change the sign of 2 z to – 2 z, then add it to 10 z, this can be written as (10 z – 2 z) and the answer will be 8 z.

Example 1:   Subtract $2x - 4y$ from $6x - 2y$

Write the polynomials as follows:

$(6x - 2y) - (2x - 4y)$

$= 6x - 2y - 2x + 4y$

Combine like terms

$= (6x - 2x) + (4y - 2y)$

$= 4x + 2y$

Example 2:   Subtract $3xy + 2y - 1$ from $6xy + 7y + 3$

Write the polynomials as follows:

$(6xy + 7y + 3) - (3xy + 2y - 1)$

$$= 6\,x\,y + 7\,y + 3 - 3\,x\,y - 2\,y + 1$$

Combine like terms

$$= (6\,x\,y - 3\,x\,y) + (7\,y - 2\,y) + (3 + 1)$$

$$= 3\,x\,y + 5\,y + 4$$

Example 3: Subtract $5\,x^2 + 2\,x$ from $9\,x^2 + 7\,x$

Write the polynomials as follows:

$$(9\,x^2 + 7\,x) - (5\,x^2 + 2\,x)$$

$$= 9\,x^2 + 7\,x - 5\,x^2 - 2\,x$$

$$= 4\,x^2 + 5\,x$$

Example 4: Subtract $2\,x^2 + x - 3$ from $7\,x^2 - 4\,x + 9$

Write the polynomials as follows:

$$(7\,x^2 - 4\,x + 9) - (2\,x^2 + x - 3)$$

$$= 7\,x^2 - 4\,x + 9 - 2\,x^2 - x + 3$$

$$= 5\,x^2 - 5\,x + 12$$

Multiplication of Monomials

To multiply two monomials use the rules of exponents explained in chapter 3.

Example 1: Multiply $(5\,x^2)(2\,x^5)$

$$= 5 \cdot 2 \cdot x^2 \cdot x^5$$

$$= 10\,x^7$$

Example 2: Multiply $(9\,y^{1/2})\,(2\,y^{1/2})$

$$= 9 \cdot 2\,y^{1/2} \cdot y^{1/2}$$

$$= 18\,y$$

Example 3:   Multiply $(7\,a^2\,b^2)\,(4\,a^4\,b^4)$

$\quad = (7\cdot4)\,(a^2\,a^4)\,(b^2\,b^4)$

$\quad = 28\,a^6\,b^6$

Example 4:   Multiply $(4\,x^2\,y^3\,z^4)\,(2\,x^3\,y^4\,z^2)$

$\quad = (4\cdot2)\,(x^2\cdot x^3)\,(y^3\cdot y^4)\,(z^4\cdot z^2)$

$\quad = 8\,x^5\,y^7\,z^6$

Multiplication of Binomials

The following are the steps to multiply two binomials:

1-   Multiply the first terms

2-   Multiply the outside terms

3-   Multiply the inside terms

4-   Multiply the last terms

5-   Add all of the above products and simplify

Example 1:   Multiply $(x-3)\,(x+3)$

$\quad = x\,(x)+x\,(3)-3(x)-3(3)$

$\quad = x^2+3x-3x-9$

$\quad = x^2-9$

Example 2:   Multiply $(y+2)\,(3y-5)$

$\quad = y\,(3\,y)+y\,(-5)+2\,(3\,y)+2\,(-5)$

$\quad = 3\,y^2-5\,y+6\,y-10$

$\quad = 3\,y^2+y-10$

38

Example 3: Multiply $(8\,a - 4)\,(4\,a - 5)$

$$= 8\,a\,(4\,a) + 8\,a\,(-5) - 4(4\,a) - 4\,(-5)$$

$$= 32\,a^2 - 40\,a - 16\,a + 20$$

$$= 32\,a^2 - 56\,a + 20$$

Example 4: Multiply $(2\,x + 4\,y)\,(3\,x - 2\,y)$

$$= (2\,x \cdot 3\,x) + 2\,x\,(-2\,y) + 4\,y\,(3\,x) + 4\,y\,(-2\,y)$$

$$= 6\,x^2 - 4\,x\,y + 12\,x\,y - 8\,y^2$$

$$= 6\,x^2 + 8\,x\,y - 8\,y^2$$

Multiplication of a Monomial by a Polynomial

Multiply the monomial by each term of the polynomial.

Example 1: Multiply y by $x^2 + 2x + 4$

$$y\,(x^2 + 2x + 4) = x^2\,y + 2x\,y + 4\,y$$

Example 2: Multiply $(y\,z)$ by $(y\,z^2 - y\,z)$

$$= y^2\,z^3 - y^2\,z^2$$

Example 3: Multiply $(a^2\,b)$ by $(b\,a^2 + 3\,a - 4\,a^2)$

$$= 3\,a^4\,b^2 + 9\,a^3\,b - 12\,a^4\,b$$

Example 4: Multiply $(2\,x\,y)$ by $(3\,x^2\,y - 2\,x\,y + x\,y^2)$

$$= 6\,x^3\,y^2 - 4\,x^2\,y^2 + 2\,x^2\,y^3$$

Multiplication of a Polynomial by a Polynomial

Multiplication of a polynomial by a polynomial is similar to multiplying a monomial by a polynomial, use the distributive law to multiply each term of the first polynomial by each term of the second polynomial, add or subtract like terms, and simplify.

Example 1:　Multiply $(y + 5)(y + 5)$

$$= y(y) + y(5) + 5(y) + 5(5)$$
$$= y^2 + 5y + 5y + 25$$
$$= y^2 + 10y + 25$$

Example 2:　Multiply $(3x + 2)(4x - 3)$

$$= 3x(4x) + 3x(-3) + 2(4x) + 2(-3)$$
$$= 12x^2 - 9x + 8x - 6$$
$$= 12x^2 - x - 6$$

Example 3:　Multiply $(a - 2)$ by $(2a^2 + a - 4)$

$$= a(2a^2) + a(a) - a(4) - 2(2a^2) - 2(a) - 2(-4)$$
$$= 2a^3 + a^2 - 4a - 4a^2 - 2a + 8$$
$$= 2a^3 + 5a^2 - 6a + 8$$

Example 4:　Multiply $(x + y + 1)$ by $(x + y - 1)$

$$= x^2 + xy + x(-1) + yx + y^2 + y(-1) + x + y + (1)(-1)$$
$$= x^2 + xy - x + xy + y^2 - y + x + y - 1$$
$$= x^2 + 2xy + y^2 - 1$$

Note that $(-x)$ plus $(+x) = 0$

and $(-y)$ plus $(+y) = 0$

SPECIAL POLYNOMIAL MULTIPLICATION

The following are special polynomial multiplication forms:

Square of a Binomial

To find the square of a binomial, square the first term, then add or subtract (depending on the signs of the terms) two times the product of the two terms, plus the square of the last term.

$$(a + b)^2 = a^2 + 2\,a\,b + b^2$$

Also

$$(a - b)^2 = a^2 - 2\,a\,b + b^2$$

Example 1: $(3\,a - 2)^2$

$$= (3\,a)^2 + 2\,(3a\,.\,-2) + (-2)^2$$

$$= 9\,a^2 - 12\,a + 4$$

Example 2: $(6 + y)^2$

$$= (6)^2 + 2\,(6\,.\,y) + (y)^2$$

$$= 36 + 12\,y + y^2$$

Example 3: $(2\,x + 4\,y)^2$

$$= (2\,x)^2 + 2\,(2\,x\,.\,4\,y) + (4\,y)^2$$

$$= 4\,x^2 + 16\,x\,y + 16\,y^2$$

Example 4: $(x - 3\,y)^2$

$$= x^2 - 2\,(x\,.\,3\,y) + (-3\,y)^2$$

$$= x^2 - 6\,x\,y + 9\,y^2$$

Difference of Two Squares

The product of two identical binomials, one of them with a plus sign between the two terms, and the other with a minus sign between the two terms equals the square of the first term minus the square of the second term.

$$(y + x)(y - x) = y^2 - x^2$$

Example 1: $(2y + x)(2y - x)$

$$= (2y)^2 - x^2$$

$$= 4y^2 - x^2$$

Example 2: $(y + 2x)(y - 2x)$

$$= y^2 - (2x)^2$$

$$= y^2 - 4x^2$$

Example 3: $(3a + 5b)(3a - 5b)$

$$= (3a)^2 - (5b)^2$$

$$= 9a^2 - 25b^2$$

Example 4: $(7y + z)(7y - z)$

$$= (7y)^2 - (z)^2$$

$$= 49y^2 - z^2$$

Division of Polynomials

Dividing a Monomial by a Monomial

To divide a monomial by a monomial the exponents quotient rule is used.

Example 1: Divide $(15\,a^3)$ by $(5\,a)$

$15\,a^3 / 5\,a = (15/5) \cdot (a^3/a)$

$= 3\,a^{3-1} = 3\,a^2$

Example 2: Divide $(36\,a^4\,b^5)$ by $(9\,a\,b^2)$

$36\,a^4\,b^5 / 9\,a\,b^2 = (36/9) \cdot (a^4/a) \cdot (b^5/b^2)$

$= 4\,a^{4-1}\,b^{5-2} = 4\,a^3\,b^3$

Example 3: Divide $(6\,x^2\,y^2)$ by $(3\,x\,y)$

$6\,x^2\,y^2 / 3\,x\,y = (6/3) \cdot (x^2/x) \cdot (y^2/y)$

$= 2x^{2-1}\,y^{2-1} = 2\,x\,y$

Example 4: Divide $(8\,x^5\,y^3)$ by $(4\,x^2\,y^2)$

$8\,x^5\,y^3 / 4\,x^2\,y^2 = (8/4) \cdot (x^5/x^2) \cdot (y^3/y^2)$

$= 2x^{5-2}\,y^{3-2} = 2\,x^3\,y$

Dividing a Polynomial by a Monomial

To divide a polynomial by a monomial divide every term of the polynomial by the monomial.

Example 1: Divide $(14x^2 + 7\,x - 2)$ by $(7\,x)$

$(14x^2 + 7\,x - 2) / (7\,x)$

$$= (14 x^2 / 7x) + (7 x / 7 x) - (2 / 7 x)$$

$$= 2x + 1 - (2 / 7x)$$

Example 2:   Divide $(12 x^3 + 6 x^2 - 3x)$ by $(3x)$

$$(12 x^3 + 9 x^2 - 3x) / (3x)$$

$$= (12 x^3 / 3x) + (9 x^2 / 3x) - (3x / 3x)$$

$$= 4 x^2 + 3 x - 1$$

Example 3:   Divide $(6 x^3 - 4 x^2 + 2x + 2)$ by $(2x)$

$$(6 x^3 - 4 x^2 + 2x + 2) / (2x)$$

$$= (6 x^3 / 2 x) - (4 x^2 / 2 x) + (2x / 2x) + (2 / 2 x)$$

$$= 3 x^2 - 2 x + 1 + (1 / x)$$

Example 4:   Divide $(4a^3 b^3 - 2 a^2 b^2 + 2 a b^3)$ by $(2 \ a \ b)$

$$4a^3 b^3 - 2 a^2 b^2 + 2 a b^3 / 2 a b$$

$$= (4a^3 b^3 / 2ab) - (2 a^2 b^2 / 2 a b) + (2 a b^3 / 2 a b)$$

$$= 2 a^2 b^2 - a b + b^2$$

Dividing a Polynomial by a Polynomial

The process of dividing a polynomial by a polynomial is similar to dividing whole numbers.

First, arrange the terms of both polynomials in a decreasing order of the exponents and write the two polynomials side by side as shown below.

Second, divide the first term of the first polynomial by the first term of the second polynomial and write the answer above the first binomial (this is the first term of the quotient).

Third, multiply the first term of the quotient by each term of the second polynomial, and write the product under the similar terms of the first polynomial.

Fourth, subtract the product from the terms of the first polynomial.

Fifth, the terms that will result from the subtraction will be considered the new first polynomial.

Repeat steps 2 – 5 above as many times as necessary to complete the division.

Example 1:   Divide $(8 y^3 + 12 y^2 + 6 y + 1)$ by $(2y + 1)$

$$
\begin{array}{r}
4 y^2 + 4 y + 1 \\
2y + 1 \overline{\smash{\big)}\ 8 y^3 + 12 y^2 + 6 y + 1} \\
8 y^3 + 4 y^2 \\
\hline
8 y^2 + 6 y + 1 \\
8 y^2 + 4 y \\
\hline
2 y + 1 \\
2 y + 1 \\
\hline
0
\end{array}
$$

Therefore, $(8 y^3 + 12 y 2 + 6 y + 1) / (2y + 1) = 4 y^2 + 4 y + 1$

Example 2:   Divide $(x^3 - 3 x^2 - x + 2)$ by $(x - 1)$

$$
\begin{array}{r}
x^2 - 2 x - 3 \\
x - 1 \overline{\smash{\big)}\ x^3 - 3 x^2 - x + 2} \\
x^3 - x^2 \\
\hline
\end{array}
$$

45

$$-2\,x^2 - x + 2$$
$$-2\,x^2 + 2\,x$$
_____

$$-3\,x + 2$$
$$-3\,x + 3$$
_____
$$-1$$

Therefore, $(x^3 - 3\,x^2 - x + 2)/(x - 1) = (x^2 - 2\,x - 3) - \dfrac{1}{x-1}$

The last term $(-\dfrac{1}{x-1})$ is called the **remainder**.

Exercise 4.1

Find the degree of the following polynomials and indicate whether they are monomial, binomial, trinomial, or a polynomial:

1-	$4\,x\,y$	2-	$9\,y\,z + 7$
3-	$3\,y + 2\,y + y$	4-	$7\,a - 5\,b + 3\,c$
5-	$10\,a + 12\,b + 6\,c - 2\,y$	6-	$11\,a\,b$
7-	$5\,y^2 + 1$	8-	$8\,x^2 + 2\,x\,y$
9-	$2\,x^2 + 3\,x - 5$	10-	$4\,x\,y^2 + 3\,x^2\,y - 2\,x\,y + 1$
11-	$16\,y^3$	12-	$3\,x^3 + 4$
13-	$6\,x^3 + 4\,x^2 - 1$	14-	$7\,y^3 - 2\,x\,y\,z + 3$
15-	$10\,a^4$	16-	$5\,a^4 + 4\,a^3$
17-	$2\,x^4 - x^2 + x$	18-	$5\,y^4 + 3\,y^3 - 2\,y^2 - 3$
19-	$12\,x^5$	20-	$11\,x^5 - 2$
21-	$3\,x^5 - 4\,x^4$	22-	$z^5 + 2\,z^4 - 3\,z^2$
23-	$9\,x^5 + 2\,x^4 - x^3 + 2$	24-	$4\,y^6$
25-	$7\,z^6 + 6\,z^4$	26-	$2\,x^6 - 4\,x^5 - 2\,x^4$
27-	$16\,x^6 - 7\,x^5 + 3\,x^3 - 7$	28-	$6\,x^6\,y^6$
29-	$3\,x^6\,y^6 + 2\,x^5\,y^4$		
30-	$15\,a^6\,b^3 - 13\,a^4\,b^2 - 11\,a^2\,b$		

Exercise 4.2

Find the sum of the following polynomials:

1- $9\,x + 3\,y,\ x + 2\,y$ 2- $7\,x - y,\ x - y$
3- $2\,z + 1,\ 3\,z + 3$ 4- $5\,y + 5\,z,\ 2\,y - 2\,z$
5- $2\,x + 4\,y,\ 3\,x - 2\,y$ 6- $4\,y - 3\,x,\ 7\,y - 2\,x$
7- $7\,y\,z + 11,\ 9 - 2\,y\,z$ 8- $3\,a\,b + 9\,c\,d,\ 7\,a\,b + c\,d$
9- $5\,y - 4\,x,\ -y - 2\,x$ 10- $6\,x + 12\,y,\ -2\,x - 8\,y$
11- $a + 2\,b - 5,\ 7\,a - b + 6$
12- $3\,x - 2\,y + 1,\ -4\,x + 2\,y - 5$
13- $7\,x - 5\,y - 3,\ 2\,x - 3\,y + 4$
14- $6\,y - 2\,x + 3,\ 2\,x - 4\,y - 7$
15- $4\,a + 2\,b,\ 2\,a - b,\ b - 2a$
16- $11\,a - 5\,b,\ a + b,\ 10\,b - 6\,a$
17- $2\,y - 3\,x,\ y + 5\,x,\ x - 3\,y$
18- $5\,x - 3\,y,\ 2\,x - 2\,y,\ 6\,y - 6\,x$
19- $4\,a\,b - 3\,b\,c + 4\,c,\ 3\,b\,c - 3\,a\,b - 2\,c,\ 7\,a\,b - 5\,c + b\,c$
20- $5\,y - 3\,x + 4,\ y + 4\,x - 7,\ 2\,x - 3\,y + 4$

Exercise 4.3

Subtract the following polynomials:

1- $3\,y - 4\,y$ 2- $2\,x - 5\,x$
3- $-5\,y + 9\,y$ 4- $-6\,y - 7\,y$
5- $9\,a\,b - 3\,a\,b$ 6- $13\,x\,y - x\,y$
7- $5\,y - 5\,y$ 8- $11\,x - 2\,x - 3\,x$
9- $7\,x + 5\,x - 6\,x$ 10- $4\,x + x + 8\,x$
11- $2\,y\,z + 4\,y\,z - y\,z$ 12- $13\,a\,b\,c - 7\,a\,b\,c - 3\,a\,b\,c$
13- $(4\,a + 6\,b) - (5\,b - 3\,a)$ 14- $(7\,a - 3\,b) - (2\,b + 6\,a)$
15- $(9\,x + 7y) + (3y - 4\,x)$
16- $(3\,y\,z + 5\,x\,y) + (3\,x\,y - 2\,y\,z)$
17- $(4\,x - 4\,y + 3) - (7\,x + y - 6)$
18- $(6\,a + 3\,b - 9) - (a + 2\,b - 1)$

19- $(3\,a - 5\,b + c) - (6\,b - a - 2\,c)$
20- $(10\,a\,b + 4\,c\,d - 3\,x\,y) - (5\,x\,y - 3\,c\,d - 9\,a\,b)$

Exercise 4.4

Multiply the following terms and simplify:

1- $2\,x^{1/2} \cdot x^{1/2}$ 2- $5\,a^2 \cdot a^2$
3- $b^2 \cdot b^3$ 4- $3\,y^3 \cdot y^5$
5- $(-2a^2)(-a^3)$ 6- $(x^4)(-y)^2$
7- $(-2\,b)^3(-b)^2$ 8- $x^3(-x^4)$
9- $(x^2\,y^3)(x\,y)$ 10- $x^2\,y\,(x^3\,y^4)$
11- $4\,x^5(x^3\,y^3)$ 12- $z^3(-5\,z\,y^2)$
13- $y^4\,z\,(-y^3\,b^3)$ 14- $-3^3\,z\,(2\,z^3\,y^5)$
15- $(b + 5)\,(b + 1)$ 16- $(3\,x - 4)\,(x - 2)$
17- $(y - 3)\,(y^2 - 5)$ 18- $(x + 1)\,(x^2 + 2\,x + 2)$
19- $(x + 4)\,(2\,x - 2)\,(2\,x + 3)$
20- $(y^2 + 2\,y - 3)\,(y^2 - 2\,y + 3)$

Exercise 4.5

Perform the following operations and simplify:

1- $(x + 4)^2$ 2- $(2\,x - 4)^2$
3- $(a + 2\,b)^2$ 4- $(2\,a + 3\,b)^2$
5- $(3\,x + y)^2$ 6- $(3y - z)^2$
7- $(2\,x - 2\,y)^2$ 8- $(7\,x\,y - 2\,y\,z)^2$
9- $(4\,x + y)\,(4\,x - y)$ 10- $(5\,b - a)\,(5\,b + a)$
11- $(3\,y\,z + 2\,x\,y)\,(3\,y\,z - 2\,x\,y)$
12- $(y - z)^2 - (y + z)^2$ 13- $(4\,y + 1)^2\,(2\,y - 2)^2$
14- $(x\,y + z)^2 - y^2 - z^2$ 15- $(2\,a\,b + c)^2 - (2\,a\,b - c)^2$
16- $(8 - 2\,a)^2\,(2 + a)^2$ 17- $(5 - 2\,b)\,(3 + b)^2$
18- $(x - 5y)^2\,(x - y)$ 19- $(y - 2\,x)^2\,(y - 2\,x)^2$
20- $(x\,y + x\,z)^2\,(x\,y - x\,z)^2$

48

Exercise 4.6

Perform the following operations and simplify:

1- $3^5 / 3^2$ 2- $4^4 / -4^3$

3- $2^5 / 2^5$ 4- $-2^3 / 2^4$

5- $-2^6 / -2^4$ 6- b^7 / b^4

7- $-c^4 / c$ 8- $(-a)^6 / (-a)^2$

9- $(y+z)^7 / (y+z)^2$ 10- $(a+b)^2 / (b+a)^5$

11- $(8\,y + 2) / 2$ 12- $(15 - 12\,z) / 3$

13- $(10\,z^4 + 4\,a\,z^2) / 2\,z^2$ 14- $(12\,y^2 - 9\,y + 6) / -3\,y$

15- $(21\,x^2 + 14\,x\,y - 7y^2) / 7\,x\,y$

16- $(x^2 + 9\,x + 14) / (x + 7)$

17- $(x^2 + 3\,x - 10) / (x + 5)$

18- $(2\,x^3 + x^2 - 21\,x + 15) / (2\,x - 5)$

19- $(2\,x^3 - 5\,x^2 + 7\,x - 6) / (2\,x - 3)$

20- $(4\,x^4 + 6\,x^3 - 16\,x^2 - 10\,x + 8) / (x^2 + x - 4)$

CHAPTER 5

FACTORING

Factoring Numbers

Multiplying two numbers such as 4 and 6 results in a product of 24. Therefore, 4 and 6 are called factors of the number 24. Other factors of 24 include the following:

1, since 1 . 24 = 24
2, since 2 . 12 = 24
3, since 3 . 8 = 24
8, since 8 . 3 = 24
12, since 12 . 2 = 24
24, since 24 . 1 = 24

Prime numbers are defined as the natural numbers greater than 1 that are divisible only by themselves and 1. Examples are the numbers 2, 3, 5, 7, 11, 13, 17, 19, 23, 29, 31, 37, 41, 43, 47, … etc.
A natural number greater than 1 that is not a prime number is called a **composite number.**

Helpful hints to factor numbers

1-   A number is divisible by 2 if it (or the last number to the right) is a multiple of 2 such as 2, 4, 6, and 8.

2-   A number is divisible by 3 if the sum of its numbers is divisible by 3.

3-   A number is divisible by 5 if the last number to the right is a 5 or a 0.

   Example 1:   the numbers 12, 24, 36, and 48 are all divisible by 2

   Example 2:   the numbers 222, 234, 246, and 258 are all divisible by 2

Example 3:   the numbers 27, 36, 45, and 54 are all divisible by 3

Example 4:   the numbers 105, 114, 123, and 132 are all divisible by 3

Example 5:   the numbers 25, 45, 65, and 85 are all divisible by 5

Example 6:   the numbers 100, 140, 180, and 290 are all divisible by 5

To find the factors of a number, check to see if the number is divisible by two, if it is find the quotient and repeat the process until you get a number that is not divisible by 2.

If the number is not divisible by 2, check to see if it is divisible by 3 or 5 or 7 and so on, continue dividing by the prime numbers until the answer is 1. The factors of this number are the divisors obtained.

Example 1:   Find the factors of 16

$16 / 2 = 8$ therefore; 2 is a factor
$8 / 2 = 4$ therefore; 2 is a factor (for the second time)
$4 / 2 = 2$ therefore; 2 is a factor (for the third time)
$2 / 2 = 1$ therefore; 2 is a factor (for the fourth time)
The factors of 16 are 2 . 2 . 2 . 2

Example 2:   Find the factors of 27

27 is not divisible by 2, try 3
$27 / 3 = 9$ therefore; 3 is a factor
$9 / 3 = 3$ therefore; 3 is a factor (for the second time)
$3 / 3 = 1$ therefore; 3 is a factor (for the third time)
The factors of 27 are 3 . 3 . 3

Example 3:   Find the factors of 120

$120 / 2 = 60$ therefore; 2 is a factor

$60 / 2 = 30$ therefore; 2 is a factor (for the second time)

30 / 2 = 15 therefore; 2 is a factor (for the third time)

The number 15 is not divisible by 2, try to divide by 3
15 / 3 = 5 therefore; 3 is a factor

Since 5 is a prime number, no additional factors for 120 will be found.

The factors of 120 are 2 . 2 . 2 . 3 . 5

Example 4:   Find the factors of 91

91 is not divisible by 2, 3, or 5

91 / 7 = 13   therefore; 7 is a factor

13 is a prime number and no additional factors for 91 will be found.

The factors of 91 are 7, 13

Common Factors

A common factor is defined as a number which is a factor of two or more other numbers. For example 5 is a common factor of 10, 15, and 20.

The Greatest Common Factor

The greatest common factor in a set of numbers is defined as the largest number that divides in each element of the set of numbers.

Example 1:   Find the greatest common factor of 14 and 28

$14 = 2 . 7$ $28 = 2 . 2 . 7$

Since only 2 and 7 are common factors of 14 and 28, then the greatest common factor of 14 and 28 is $2 . 7 = 14$.

Example 2:   Find the greatest common factor of 10 and 30

$10 = 2 . 5$ $30 = 2 . 3 . 5$

Since only 2 and 5 are common factors of 10 and 30, then the greatest common factor of 10 and 30 is $2 . 5 = 10$.

Example 3:   Find the greatest common factor of 27 and 54

$27 = 3 . 3 . 3$ $54 = 2 . 3 . 3 . 3$

Since only 3 is a common factor of 27 and 54, then the greatest common factor of 27 and 54 is $3 . 3 . 3 = 27$.

Example 4:   Find the greatest common factor of 210 and 315

$210 = 2 . 3 . 5 . 7$ $315 = 3 . 3 . 5 . 7$

Since only 3, 5, and 7 are common factors of 210 and 315, then the greatest common factor of 210 and 315 is $3 . 5 . 7 = 105$.

The greatest common factor is helpful in **reducing fractions** with large numbers to an equivalent fraction with smaller numbers.

Example 1:   Reduce the fraction $\dfrac{72}{108}$

$72 = 2 . 2 . 2 . 3 . 3$
$108 = .2 . 2 . 3 . 3 . 3$

The greatest common factor is $2 \cdot 2 \cdot 3 \cdot 3 = 36$

$$\frac{72}{108} = \frac{36.2}{36.3} = \frac{2}{3}$$

Example 2: Reduce the fraction $\dfrac{110}{154}$

$110 = 2 \cdot 5 \cdot 11$
$154 = 2 \cdot 7 \cdot 11$

The greatest common factor is $2 \cdot 11 = 22$

$$\frac{110}{154} = \frac{22.5}{22.7} = \frac{5}{7}$$

Factoring Polynomials

The greatest common factor in polynomials can be found by factoring the integers into prime factors, writing the factors in the exponential form, then multiplying the common bases with the lowest exponent.

Example 1: Find the greatest common factor of $25\ x^5 y^3$, $125\ x^4 y$, $625\ x^3 y^2$

Factor each term:

$$25\ x^5 y^3 = 5^2 \cdot x^5 \cdot y^3$$

$$125\ x^4 y = 5^3 \cdot x^4 \cdot y$$

$$625\ x^3 y^2 = 5^4 \cdot x^3 \cdot y^2$$

It is clear from the above analysis that the common bases are 5, x, y

54

The least exponents of these common bases are 2, 3, and 1, therefore the greatest common factor is
$$5^2 \; x^3 \; y = 25 \; x^3 \; y$$

Example 2: Find the greatest common factor of $9 \, y^3 \, (x-1)$, $27 \, y^5 \, (x-1)$, $81 \, y^7 \, (x-1)$.

Factor each term:

$$9 \, y^3 \, (x-1) = 3^2 \cdot y^3 \, (x-1)$$

$$27 \, y^5 \, (x-1) = 3^3 \cdot y^5 \, (x-1)$$

$$81 \, y^7 \, (x-1) = 3^4 \cdot y^7 \, (x-1)$$

The greatest common factor $= 3^2 \; y^3 \, (x-1)$

Example 3: Find the greatest common factor of $8 \, b^3 \, (a-4)^3$, $16 \, b^4 \, (a-4)^2$, $4 \, b^2 \, (a-4)$.

Factor each term:

$$8 \, b^3 \, (a-4)^3 = 2^3 \cdot b^3 \, (a-4)^3$$

$$16 \, b^4 \, (a-4)^2 = 2^4 \cdot b^4 \, (a-4)^2$$

$$4 \, b^2 \, (a-4) = 2^2 \cdot b^2 \, (a-4)$$

The greatest common factor $= 2^2 \; b^2 \, (a-4)$

Example 4: Find the greatest common factor of $36 \, x^2 y^3 z$, $45 \, x^3 \, y^4 z^2$, $81 \, x \, y^2 z^3$.

Factor each term:

$$36\,x^2y^3z = 2^2 \cdot 3^2 \cdot x^2y^3z$$

$$45\,x^3\,y^4\,z^2 = 5 \cdot 3^2 \cdot x^3\,y^4\,z^2$$

$$81\,x\,y^2z^3 = 3^2 \cdot 3^2 \cdot x\,y^2z^3$$

The greatest common factor $= 3^2\ x\ y^2\ z$

Factoring Binomials

Factoring the difference between two squares

The binomial $(a^2 - b^2)$ is an example of the difference between two squares. This binomial can be factored to $(a + b)\,(a - b)$. The first factor is the square root of the first term plus the square root of the second term, and the second factor is the square root of the first term minus the square root of the second term.

Example 1:   Factor $16\,x^2 - 169$

The square root of $16 = 4$, the square root of $x^2 = x$. and the square root of $169 = 13$

Therefore, $16\ x^2 - 169 = (4x + 13)\,(4x - 13)$

Example 2:   Factor $81\,y^4 - x^4$

$$81\,y^4 - x^4 = (9\,y^2 + x^2)\,(9\,y^2 - x^2)$$

This is not the final answer because the term $(9\,y^2 - x^2)$ is also difference between two squares, therefore we will continue to factor this term.

$$9 y^2 - x^2 = (3 y + x)(3 y - x)$$

The final answer is $(9 y^2 + x^2)(3 y + x)(3 y - x)$

Example 3: Factor $49 z^6 - 16 z^4$

The square root of $49 = 7$, the square root of $z^6 = z^3$, the square root of $16 = 4$, and the square root of $z^4 = z^2$

Therefore, $49 z^6 - 16 z^4 = (7z^3 + 4z^2)(7z^3 - 4z^2)$

Example 4: Factor $4 x^2 y^2 - 9 z^2$

The square root of $4 = 2$, the square root of $x^2 = x$, the square root of $y^2 = y$, the square root of $9 = 3$, and the square root of $z^2 = z$

Therefore, $4 x^2 y^2 - 9 z^2 = (2 x y + 3z)(2 x y - 3z)$

Factoring the Sum of Two Cubes

Factoring the sum of two cubes such as $y^3 + x^3$ will yield two factors. The first factor consists of two terms. The first term of the first factor is the cubic root of the first term (cubic root of y^3), this equals y, and the second term is the cubic root of the second term (cubic root of x^3), this equals x.
This means that the first factor is $(y + x)$.

The second factor will include three terms, these are, the square of the first term in the first factor, this equal y^2. The second term equals the negative product of the two terms in the first factor, this equals $(- y x)$. The third term of the second factor is the square of the second term in the first factor, this equals (x^2).

Following the above steps will yield $(y + x)(y^2 - x y + x^2)$.

Example 1:   Factor $(x^3 + 27)$

The first factor $= (x + 3)$

The second factor $= (x^2 - 3x + 9)$

$x^3 + 27 = (x + 3)(x^2 - 3x + 9)$

Example 2:   Factor $(27x^3 + 64y^3)$

The first factor $= (3x + 4y)$

The second factor $= (9x^2 - 12x + 16y^2)$

$27x^3 + 64y^3 = (3x + 4y)(9x^2 - 12xy + 16y^2)$

Example 3:   Factor $(8x^3 + 125y^6)$

The first factor $= (2x + 5y^2)$

The second factor $= (4x^2 - 10xy^2 + 25y^4)$

$8x^3 + 125y^6 = (2x + 5y^2)(4x^2 - 10xy^2 + 25y^4)$

Example 4:   Factor $(128x^3 + 16y^3)$
$(128x^3 + 16y^3) = 2(64x^3 + 8y^3)$

Ignore the factor 2 for now.

The first factor $= (4x + 2y)$

The second factor $= (16x^2 - 8xy + 4y^2)$

$(128x^3 + 16y^3) = 2(4x + 2y)(16x^2 - 8xy + 4y^2)$

Note that the factor 2 that was left out is multiplied by the two factors.

Factoring the Difference Between Two Cubes

Factoring the difference between two cubes such as $y^3 - x^3$ will yield two factors. The first factor consists of two terms. The first term of the first factor is the cubic root of the first term (cubic root of y^3), this equals y, and the second term is the cubic root of the second term (cubic root of x^3), this equals x. This means that the first factor is (y – x).

The second factor will include three terms, these are, the square of the first term in the first factor, this equal y^2. The second term equals the negative product of the two terms in the first factor, this equals (y x). The third term of the second factor is the square of the second term in the first factor, this equals (x^2). Following the above steps will yield (y – x) (y^2 + x y + x^2).

Example 1:   Factor $y^3 - 8x^3$

The first factor = (y – 2x)

The second factor = (y^2 + 2x y + 4x^2)

$y^3 - 8x^3$ = (y – 2x) (y^2 + 2x y + 4x^2)

Example 2:   Factor $64x^3 - 8$

The first factor = (4x – 2)

The second factor = (16x^2 + 8x + 4)

$64x^3 - 8$ = (4x – 2) (16x^2 + 8x + 4)

Example 3:   Factor $27a^3 - 125b^3$

The first factor = (3 a – 5 b)

The second factor = $(9\,a^2 + 15\,a\,b + 25\,b^2)$

$27\,a^3 - 125\,b^3 = (3\,a - 5\,b)\,(9\,a^2 + 15\,a\,b + 25\,b^2)$

Example 4:   Factor $[64\,x^3 - (y+1)^3]$

The first factor = $[4\,x - (y+1)]$

The second factor = $[16\,x^2 + 4\,x\,(y+1) + (y+1)^2]$

$[64\,x^3 - (y+1)^3] =$
$[4\,x - (y+1)]\,[16\,x^2 + 4\,x\,(y+1) + (y+1)^2]$

Factoring Trinomials

Factoring a trinomial such as $x^2 - 12\,x + 35$ results in two factors, and each one of the factors includes two terms, such as $(x-5)\,(x-7)$.

To find the two factors of a trinomial, follow the following steps:

1- To find the first term of the first factor and the first term of the second factor (which is x in the above example), take the square root of the first term in the trinomial (square root of x^2).

2- To find the second term of the first factor (which is -5 in the above example) as well as the second term of the second factor (which is -7 in the above example), find two numbers whose product equals the third term of the trinomial (which is 35).

3- The product of the two numbers chosen ($-5\,x$ and -7) must equal the third term in the trinomial (35), and the sum of the two numbers chosen, taking the signs into consideration $[(-5 + (-7)]$, must equal the coefficient of the second term of the trinomial (-12).

60

Example 1:   Factor $y^2 + 8y + 15$

The first term of the first and second factors is equal to the square root of $y^2 = y$

Factor 15 into two numbers such that the product of the two numbers = 15, and the sum of the two numbers = 8

The two numbers are 5 and 3

$$y^2 + 8y + 15 = (y + 5)(y + 3)$$

Example 2:   Factor $y^2 + 2y - 8$

The first term of the first and second factors is equal to the square root of $y^2 = y$

Factor - 8 into two numbers such that the product of the two numbers = – 8, and the sum of the two numbers = 2

The two numbers are 4 and – 2

$$y^2 + 2y - 8 = (y + 4)(y - 2)$$

Example 3:   Factor $x^2 - 12x + 20$

The first term of the first and second factors is equal to the square root of $x^2 = x$

Factor 20 into two numbers such that the product of the two numbers = 20, and the sum of the two numbers = -12

The two numbers are – 10 and – 2

$$x^2 - 12x + 20 = (x - 10)(x - 2)$$

Example 4:   Factor $x^2 - 5x - 14$

The first term of the first and second factors is equal to the square root of $x^2 = x$

Factor 14 into two numbers such that the product of the two numbers $= -14$, and the sum of the two numbers $= -5$

The two numbers are -7 and 2

$$x^2 - 5x - 14 = (x - 7)(x + 2)$$

Factoring by Grouping

Factoring polynomials with four terms can sometimes be completed by grouping three terms together leaving only one term, or by grouping each two terms together.

Grouping Two Terms

We may be able to factor some polynomials with four terms by grouping every two terms together. If the two chosen terms do not have a common factor, try grouping another two terms.

Example 1:   Factor $y^3 + 2y^2 + 2y + 4$

Try grouping the first two terms

$(y^3 + 2y^2) + (2y + 4)$

Factor y^2 from the first two terms, and 2 from the other two terms.

$$(y^3 + 2y^2) + (2y + 4) = y^2(y + 2) + 2(y + 2)$$

Factor $(y + 2)$

$$y^2(y + 2) + 2(y + 2) = (y + 2)(y^2 + 2)$$

Example 2:   Factor $y^3 + 3y^2 - 9y - 27$

Try grouping the first two terms

$$(y^3 + 3y^2) - (9y + 27)$$

Factor y^2 from the first two terms, and 9 from the other two terms.

$$(y^3 + 3y^2) - (9y + 27) = y^2(y + 3) - 9(y + 3)$$

Factor $(y + 3)$

$$y^2(y + 3) - 9(y + 3) = (y + 3)(y^2 - 9)$$

The term $(y^2 - 9)$ is the difference between two squares and can be factored to $(y + 3)(y - 3)$

The final answer is $(y + 3)(y + 3)(y - 3)$
Which can also be written as $(y + 3)^2(y - 3)$

Example 3:   Factor $4y^3 + 6y^2 + 4y + 6$

Try grouping the first two terms

$$(4y^3 + 6y^2) + (4y + 6)$$

Factor $2y^2$ from the first two terms, and 2 from the other two terms.

$$(4y^3 + 6y^2) + (4y + 6) = 2y^2(2y + 3) + 2(2y + 3)$$

Factor $(2y + 3)$

$$2y^2(2y + 3) + 2(2y + 3) = (2y + 3)(2y^2 + 2)$$

Example 4:   Factor $y^3 + y^2 - 3y - 27$

Try to group the first two terms $(y^3 + y^2)$

$$y^3 + y^2 = y^2(y + 1)$$

Try to group the second two terms $(-3y - 27)$

$$(-3y - 27) = -3(y + 9)$$

As you can see there are no common factors between the two groups.

Try grouping the first term and the fourth term

$$(y^3 - 27) + (y^2 - 3y)$$

The first group $(y^3 - 27)$ is a difference between two cubes, and can be factored as follows:

$$(y^3 - 27) = (y - 3)(y^2 + 3y + 9)$$

Factor the second group $(y^2 - 3y)$

$$(y^2 - 3y) = y(y - 3)$$
Note that $(y - 3)$ is a common factor between the two groups.

$$(y^3 - 27) + (y^2 - 3y) = (y - 3)(y^2 + 3y + 9) + y(y - 3)$$

$$= (y - 3)[(y^2 + 3y + 9) + y]$$

$$= (y - 3)(y^2 + 3y + 9 + y)$$

$$= (y - 3)(y^2 + 4y + 9)$$

Grouping Three Terms

Try grouping two of the three terms that are squared along with the term that is not squared. For example $a^2 - 4b^2 + 4c^2 - 4ac$ try grouping $(a^2 - 4ac + 4c^2)$ and leave the fourth term $(-4b^2)$ separately.

The next step is to see if the chosen three terms can be factored as a squared term. For example can the three grouped terms $(a^2 - 4ac + 4c^2)$ be factored as a square. In this example the answer is yes, $(a^2 - 4ac + 4c^2) = (a - 2c)^2$.

The next step is to rewrite the polynomial in a factored form including factoring the factored square $(a - 2c)^2$ and the fourth term of the original polynomial $(-4b^2)$.

$$a^2 - 4b^2 + 4c^2 - 4ac = (a^2 - 4ac + 4c^2) - 4b^2$$

$$= (a - 2c)^2 - 4b^2$$

$$= [(a - 2c) + 2b][(a - 2c) - 2b]$$

$$= (a - 2c + 2b)(a - 2c - 2b)$$

Example 1:   Factor $a^2 - b^2 - 8b - 16$

Try to group $(-b^2 - 8b - 16)$, and leave a^2 separate

$$a^2 - b^2 - 8b - 16 = a^2 - (b^2 + 8b + 16)$$

$$= a^2 - (b + 4)^2$$

$$= [a + (b + 4)][a - (b + 4)]$$

$$= (a + b + 4)(a - b - 4)$$

Example 2: Factor $4a^2 - b^2 - 9c^2 + 6bc$

Try to group $(-b^2 - 9c^2 + 6bc)$ and leave $4a^2$ separate

$4a^2 - b^2 - 9c^2 + 6bc = 4a^2 - (b^2 - 6bc + 9c^2)$

$= [2a + (b - 3c)][2a - (b - 3c)]$

$= (2a + b - 3c)(2a - b + 3c)$

Example 3: Factor $x^3 + 2x^2y + xy^2 - 25x$

Factor x from all the terms

$x^3 + 2x^2y + xy^2 - 25x = x(x^2 + 2xy + y^2 - 25)$

Try grouping the first three terms

$x[(x^2 + 2xy + y^2) - 25)]$

$= x[(x + y)^2 - 25]$

$= x(x + y + 5)(x + y - 5)$

Example 4: Factor $4x^2 - 12xy + 9y^2 + 25z^2$

Try to group $(4x^2 - 12xy + 9y^2)$ and leave $25z^2$ separate

$4x^2 - 12xy + 9y^2 + 25z^2$

$= (4x^2 - 12xy + 9y^2) + 25z^2$

$= (2x - 3y)^2 + 25z^2$

$= [(2x - 3y) + 5z][(2x - 3y) + 5z]$

$= (2x - 3y + 5z)(2x - 3y + 5z)$

Exercise 5.1

Find the factors of the following numbers:

1-	64	2-	81
3-	100	4-	120
5-	144	6-	225
7-	240	8-	273
9-	288	10-	360
11-	363	12-	420
13-	432	14-	500
15-	507	16-	630
17-	675	18-	686
19-	720	20-	840

Exercise 5.2

Find the greatest common factor of the following numbers:

1-	9, 12, 15	2-	5, 15, 30
3-	10, 30, 50	4-	12, 27, 48
5-	14, 18, 22	6-	16, 20, 24
7-	21, 42, 63	8-	26, 39, 52
9-	24, 30, 36	10-	24, 54, 96
11-	6, 36, 216	12-	16, 36, 64
13-	18, 36, 54	14-	18, 54, 162
15-	35, 42, 49	16-	42, 54, 60
17-	22, 44, 132	18-	25, 75, 250
19-	42, 63, 147	20-	66, 99, 165

Exercise 5.3

Find the greatest common factor of the following terms:

1-	$a^2 b^2, a^3 b^3, a^4 b^4$	2-	$x^2 y^3, x^3 y^2, x^4 y$
3-	$y^4 z^2, y^2 z^3, y^3 z$	4-	$9(a+1), 18(a+1)^2$

5-	$28\,(x+3)^{2}$, $56\,(x+3)$		6-	$z\,(z-3)$, $z^{2}\,(3-z)$
7-	$3\,a^{2}\,(a-5)$, $12\,a\,(5-a)$			
8-	$2(4\,x-3)$, $8\,(4\,x-3)^{2}$			
9-	$5\,b^{3}\,(c+11)$, $20\,b^{2}\,(2c+22)$			
10-	$(y+3)\,(y-4)$, $(5+y)\,(4-y)$			
11-	$(x-5)\,(3\,x+2)$, $(5-x)$, $(3\,x+3)$			
12-	$x^{3}\,y$, $x^{3}\,y^{2}$, $x^{4}\,y^{3}$		13-	$a^{3}\,b^{3}$, $a^{3}\,b^{2}$, $a^{3}\,b^{4}$
14-	$12(y-2)$, $48(y-2)$		15-	$14(z+5)$, $49(z+5)$
16-	$22\,(z+7)$, $33\,(2z+14)$		17-	$6(2\,x-8)$, $6(3\,x-12)$
18-	$16(5\,a-3)$, $32(3-5\,a)$			
19-	$25(5+x)^{2}$, $125(5+x)^{3}$			
20-	$(y-7)\,(5y+3)$, $(7-y)\,(10\,y+6)$			

Exercise 5.4

Factor the following terms by finding the common factors:

1-	$5\,y+10$		2-	$21\,y+28$
3-	$3\,x-9$		4-	$72-6\,x$
5-	$2\,a-6\,a^{2}$		6-	$4\,a\,b-8\,b$
7-	$3\,a\,b+a^{2}$		8-	$5\,a^{2}-10\,a^{2}\,b$
9-	$12\,x\,y^{2}+4\,y^{3}$		10-	$6\,x^{3}-3\,x^{2}$
11-	$7\,x^{2}\,y^{2}-49\,x^{3}y^{3}$		12-	$(y+2)^{2}+(y+2)$
13-	$4\,(y-3)^{2}+8(y-3)$			
14-	$6(x+5)^{2}-3(2\,x+10)$			
15-	$7\,y^{3}-14\,y^{3}\,z^{3}+28\,y^{3}\,z^{3}$			
16-	$8\,a^{2}-16\,a^{2}\,b+24\,a^{2}\,b^{2}$			
17-	$9\,y^{2}\,z^{2}+27\,y^{2}\,z^{3}-81\,y^{2}\,z^{4}$			
18-	$14\,x^{3}y^{3}+28\,x^{3}\,y^{2}-42\,x^{3}\,y$			
19-	$5\,x\,y^{2}\,z^{2}+10\,x\,y^{3}\,z^{3}+15\,x\,y^{2}\,z^{2}$			
20-	$6\,x^{2}\,y^{3}\,z^{4}+12\,x^{2}\,y^{4}\,z^{2}+18\,x^{3}\,y^{2}\,z^{4}$			

Exercise 5.5

Factor the following terms using the difference between two squares rule, the sum of two cubes rule, or the difference between cubes rule, whichever applies:

1- $y^2 - 16$ 2- $y^2 - 49$

3- $a^2 + 81$ 4- $a^2 + 144$

5- $36 - z^2$ 6- $64z^2 - 4$

7- $16x^2 - 25y^2$ 8- $x^2y^2 + 4$

9- $3a^4 - 12a^2b^2$ 10- $100a^4 - 25b^4$

11- $49 - 16y^4$ 12- $36y^4 - 16z^4$

13- $x^3 - 64$ 14- $x^3 - 216$

15- $x^3 - 1$ 16- $1 - 27x^3$

17- $8y^3 - z^3$ 18- $2x^3 - 54$

19- $4y^3 + 32$ 20- $y^3 + 1$

21- $(z+1)^3 + y^3$ 22- $108x^3 + 32y^3$

Exercise 5.6

Factor the following trinomials:

1- $x^2 - 3x + 2$ 2- $x^2 + x - 6$

3- $x^2 + 4x + 3$ 4- $x^2 - x - 6$

5- $x^2 + 7x + 12$ 6- $x^2 + x - 12$

7- $x^2 - 3x - 10$ 8- $x^2 + 2x - 15$

9- $x^2 - 6y + 5y^2$ 10- $x^2 + xy - 2y^2$

11- $3x^2 - 27x + 42$ 12- $3x^2 + 6xy + 3y^2$

13- $2x^2 - 16x + 30$ 14- $5x^2 + 10xy - 15y^2$

15- $4x^2 + 4x - 63$ 16- $18x^2 + 72x + 70$

17- $6x^2 + 6xy - 72y^2$ 18- $12x^2 - 12xy - 105y^2$

19- $20x^2 - 80xy + 75y^2$

20- $84x^2 + 77xy + 14y^2$

Exercise 5.7

Factor the following terms by grouping:

1- $\quad 9xy - 21x - 3yz + 7z$

2- $\quad 28ax - 14bx - 14ay + 7by$

3- $\quad 7x^2 - 21x - xy + 3y$ $\qquad$ 4- $\quad y^2 - x^2 + 6xz - 9z^2$

5- $\quad y^2 - z^2 - 6z - 9$ $\qquad$ 6- $\quad x^2 - y^2 - 5z - 10$

7- $\quad a^2 + 4ab + 4b^2 - 16c^2$ $\qquad$ 8- $\quad x^2 - 6xy + 9y^2 - 25z^2$

9- $\quad x^2 - 4y^2 + 2x + 1$ $\qquad$ 10- $\quad y^2 + 4z^2 - 4yz - 4$

11- $\quad 4a^2 + b^2 + 4ab - 25$ $\qquad$ 12- $\quad 2a^2 + 2b^2 - 4ab - 18$

13- $\quad 9x^2 + 18xy + 9y^2 - 16$ $\qquad$ 14- $\quad 5x^2 - 10xy + 5y^2 - 45$

15- $\quad x^3 + x^2 + 5x + 5$ $\qquad$ 16- $\quad y^3 - y^2 - 3y + 3$

17- $\quad 2x^3 + x^2 + 6x + 3$ $\qquad$ 18- $\quad 9y^3 - 27y^2 - y + 3$

19- $\quad y^3 + y^2 - 2y - 8$ $\qquad$ 20- $\quad 8a^3 + b^3 - 6a - 3b$

21- $\quad a^3 + a^2 + 2a + 2$ $\qquad$ 22- $\quad y^3 + y^2 + 2y + 2$

23- $\quad x^3 + 2x^2y + xy^2 - 25x$

24- $\quad a^3 + 2a^2 - 4a - 8$

CHAPTER 6

RADICALS

Definitions

This symbol $\sqrt{}$ is called the radical sign

In the term $\sqrt[n]{y}$ n is called the index and it should be a positive integer greater than 1.
y is called the **radicand**.
The whole term is read as the **nth root** of y.

The term $\sqrt[n]{y}$ can also be written as $y^{1/n}$

The index n (the number) must be written for all roots except for **square root**.
This means that instead of writing $\sqrt[2]{y}$, we always write it as $\sqrt{y}$.

Example 1:   Find the square root of 16
This is written as $\sqrt{16}$, and it means what is the number you multiply by itself to get a product of 16?
The answer is 4, because $(4 \cdot 4) = 16$

Example 2:   Find the **cubic root** of 8
This is written as $\sqrt[3]{8}$, and it means what is the number you multiply by itself three times to get a product of 8?
The answer is 2, because $(2 \cdot 2 \cdot 2) = 8$

Example 3:   Find the fourth root of 81, we write it as ($\sqrt[4]{81}$)

$81 = 3 \cdot 3 \cdot 3 \cdot 3$
The answer is 3, because $(3 \cdot 3 \cdot 3 \cdot 3) = 81$

Example 4:   Find the fifth root of 32, we write it as ($\sqrt[5]{32}$)

$$32 = 2 \cdot 2 \cdot 2 \cdot 2 \cdot 2$$
The answer is 2, because $(2 \cdot 2 \cdot 2 \cdot 2 \cdot 2) = 32$

Obviously, it is not always easy to find the nth root of a number just by looking at the term. The nth root of any number can be easily found by using a computer or a scientific calculator.

A number is called a **perfect square** when its square root is a rational number. Examples of perfect squares are the numbers 1, 4, 9, 16, 25

$$\sqrt{1} = 1 \qquad \sqrt{4} = 2 \qquad \sqrt{9} = 3 \qquad \sqrt{16} = 4 \qquad \sqrt{25} = 5$$

Radicals can be expressed as fractional exponents.

Example 1: $\sqrt{25} = 25^{1/2}$

Example 2: $\sqrt[3]{7} = 7^{1/3}$

Example 3: $\sqrt[4]{y-3} = (y-3)^{1/4}$

Example 4: $\sqrt[5]{x^2} = x^{2/5}$

Example 5: $\sqrt{a^2 + b^2} = (a^2 + b^2)^{1/2}$

Forms of Radical Expressions:

Radicals with a Negative Radicand

If the index of a radical expression is an even number and the radicand is a negative number, the answer is not a real number, it is an **imaginary number** and is indicated by using the symbol (***i***).

72

The imaginary number (i) = $\sqrt{-1}$, and $i^2 = -1$

Example 1: $\sqrt{-16} = \sqrt{16}.\sqrt{-1} = 4\,i$
The answer is not a real number, it is an imaginary number.

Example 2: $\sqrt{-x^2 y^2}$, where x and y are greater than zero.
$= \sqrt{x^2 y^2}.\sqrt{-1} = x\,y\,i$
The answer is not a real number, it is an imaginary number.

Example 3: $\sqrt[4]{-81} = \sqrt[4]{81}.\sqrt[4]{-1} = 3\,i$
The answer is not a real number, it is an imaginary number.

Example 4: $\sqrt[4]{-y^4}$, where y is greater than zero.
$= \sqrt[4]{y^4}.\sqrt[4]{-1} = y\,i$
The answer is not a real number, it is an imaginary number

If the index of a radical expression is an odd number and the radicand is a negative number, the answer is the negative of the radical expression, and changing the sign of the radicand from negative to positive.

Example 1: $\sqrt[3]{-y}$, where y is greater than zero.

$= -\sqrt[3]{y}$

Example 2: $\sqrt[3]{-y^3}$, where y is greater than zero.

$= -\sqrt[3]{y^3} = -y$

Example 3: $\sqrt[3]{-125}$

$$= -\sqrt[3]{125} = -\sqrt[3]{5^3} = -5$$

Example 4: $\sqrt[5]{-243}$

$$= -\sqrt[5]{243} = -\sqrt[5]{3^5} = -(3)^{5/5} = -3$$

Radicals with Common Factors

When all the radical exponents and the index have a common factor, to simplify the radical, divide all the radical exponents and the index by the common factor.

Example 1: $\sqrt[4]{x^2 y^2} = \sqrt{xy}$

Example 2: $\sqrt[8]{x^2 y^4} = \sqrt[4]{xy^2}$

Example 3: $\sqrt[6]{2^3 \cdot 3^3} = \sqrt[2]{2 \cdot 3} = \sqrt{6}$

Example 4: $\sqrt[6]{x^3 y^6 z^3} = \sqrt{xy^2 z}$

Example 5: $\sqrt[4]{\dfrac{x^2 y^2}{a^4 b^4}} = \sqrt{\dfrac{xy}{a^2 b^2}} = \dfrac{\sqrt{xy}}{ab}$

Separating Factors

When the radical exponents of some factors are not multiples of the index, these factors can be separated into two factors such that one of these factors is a multiple of the index. To simplify the newly created radical, write the factors that are multiples of the index under a radical sign, and the other factors under another radical sign.

Example 1:　　$\sqrt[4]{x^9}$　$=$　$\sqrt[4]{x^8 . x}$

$$= \sqrt[4]{x^8} . \sqrt[4]{x} = x^2 . \sqrt[4]{x}$$

Example 2:　　$\sqrt[3]{8y^4}$　$=$　$\sqrt[3]{2^3 . y^3 . y}$

$$= \sqrt[3]{2^3 y^3} . \sqrt[3]{y} = 2y . \sqrt[3]{y}$$

Example 3:　　$\sqrt[3]{16y^4 z^6}$

$$= \sqrt[3]{(2^3 . 2)(y^3 . y)z^6}$$

$$= \sqrt[3]{2^3 y^3 z^6} . \sqrt[3]{2y}$$

$$= 2 y z^2 . \sqrt[3]{2y}$$

Example 4:　　$\sqrt[6]{x^4 y^4 z^5}$　$=$　$\sqrt[6]{(x^3 ..x)(y^3 .y)(z^3 .z^2)}$

$$= \sqrt[6]{x^3 y^3 z^3} . \sqrt[6]{xyz^2}$$

$$= \sqrt{xyz} . \sqrt[6]{xyz^2}$$

Combining Similar Radical Expressions

Similar radical expressions are defined as radical expressions with the same radical index and radicand. For example $2\sqrt{2}$ and $6\sqrt{2}$ are similar radical expressions.

Some radical expressions may not seem to be similar when you first look at them, but when simplified it will be clear that they are similar.

Example 1: $\sqrt{8}$ and $\sqrt{18}$

$$\sqrt{8} = \sqrt{4.2} = \sqrt{2^2.2} = 2\sqrt{2}$$

$$\sqrt{18} = \sqrt{9.2} = \sqrt{3^2.2} = 3\sqrt{2}$$

Example 2: $\sqrt{45}$ and $\sqrt{80}$

$$\sqrt{45} = \sqrt{9.5} = \sqrt{3^2.5} = 3\sqrt{5}$$

$$\sqrt{80} = \sqrt{16.5} = \sqrt{4^2.5} = 4\sqrt{5}$$

Example 3: $\sqrt[3]{24}$ and $\sqrt[3]{81}$

$$\sqrt[3]{24} = \sqrt[3]{8.3} = \sqrt[3]{2^3.3} = 2\sqrt[3]{3}$$

$$\sqrt[3]{81} = \sqrt[3]{27.3} = \sqrt[3]{3^3.3} = 3\sqrt[3]{3}$$

Example 4: $\sqrt[4]{32}$ and $\sqrt[4]{162}$

$$\sqrt[4]{32} = \sqrt[4]{16.2} = \sqrt[4]{2^4 .2} = 2\sqrt[4]{2}$$

$$\sqrt[4]{162} = \sqrt[4]{81.2} = \sqrt[4]{3^4 .2} = 3\sqrt[4]{2}$$

Similar radical expressions can be simplified by combining them.

Example 1: Simplify $\sqrt{50} + \sqrt{8}$

$$\sqrt{50} + \sqrt{8} = \sqrt{25.2} + \sqrt{4.2}$$

$$= \sqrt{5^2 .2} + \sqrt{2^2 .2} = 5\sqrt{2} + 2\sqrt{2} = 7\sqrt{2}$$

Example 2: Simplify $\sqrt{45} - \sqrt{20}$

$$\sqrt{45} - \sqrt{20} = \sqrt{9.5} - \sqrt{4.5}$$

$$= \sqrt{3^2 .5} - \sqrt{2^2 .5} = 3\sqrt{5} - 2\sqrt{5} = \sqrt{5}$$

Example 3: Simplify $\sqrt{243} - \sqrt{27}$

$$\sqrt{243} - \sqrt{27} = \sqrt{81.3} - \sqrt{9.3}$$

$$= \sqrt{9^2 .3} - \sqrt{3^2 .3}$$

$$= 9\sqrt{3} - 3\sqrt{3} = 6\sqrt{3}$$

Example 4   Simplify $\sqrt{176}\ +\ \sqrt{99}\ -\ \sqrt{44}$

$$\sqrt{176}\ =\ \sqrt{16.11}\ =\ \sqrt{4^2.11}\ =\ 4\sqrt{11}$$

$$\sqrt{99}\ =\ \sqrt{9.11}\ =\ \sqrt{3^2.11}\ =\ 3\sqrt{11}$$

$$\sqrt{44}\ =\ \sqrt{4.11}\ =\ \ =\ 2\sqrt{11}$$

$$4\sqrt{11}\ +\ 3\sqrt{11}\ -\ 2\sqrt{11}\ =\ 5\sqrt{11}$$

Adding and Subtracting Similar Radicals

Radical expressions with the same radical index and radicand are called similar radicals. Only similar radicals can be added or subtracted.

Example 1: $7\sqrt{5} + 3\sqrt{5}$

$$= (7+3)\sqrt{5} = 10\sqrt{5}$$

Example 2: $9\sqrt{11}\ +\ 11\sqrt{11}$

$$= (9+11)\sqrt{11} = 20\sqrt{11}$$

Example 3: $17\sqrt{7}\ -\ 6\sqrt{7}\ -\ 3\sqrt{7}$

$$= (17-6-3)\sqrt{7}\ =\ 8\sqrt{7}$$

Example 4: $5\sqrt{3}\ +\ \sqrt{12}$

$$= 5\sqrt{3}\ +\ \sqrt{4.3}\ =\ 5\sqrt{3}\ +\ \sqrt{2^2.3}$$

$$= 5\sqrt{3}\ +\ 2\sqrt{3}\ =\ 7\sqrt{3}$$

Example 5: $3 \sqrt[3]{54} - 2 \sqrt[3]{16}$

$$= 3 \sqrt[3]{27.2} - 2 \sqrt[3]{8.2}$$

$$= 3 \sqrt[3]{3^3 .2} - 2 \sqrt[3]{2^3 .2}$$

$$= (3 . 3) \sqrt[3]{2} - (2 . 2) \sqrt[3]{2}$$

$$= 9 \sqrt[3]{2} - 4 \sqrt[3]{2} = 5 \sqrt[3]{2}$$

Multiplying Radicals

Product Rule

To multiply radicals that have the same index, write the radical sign with the index and multiply the radicands under the radical sign.

$\sqrt[n]{x} . \sqrt[n]{y} = \sqrt[n]{xy}$ provided n is a positive integer, and x and y are real numbers.

Example 1: $\sqrt{2} . \sqrt{11}$
$$= \sqrt{2.11} = \sqrt{22}$$

Example 2: $2 \sqrt{8} . 5 \sqrt{18}$
$$= (2 . 5) \sqrt{8.18} = 10 \sqrt{144}$$
$$= 10 . 12 = 120$$

Example 3: $3\sqrt[3]{7} \cdot 5\sqrt[3]{6}$

$= (3 \cdot 5)\sqrt[3]{7.6} = 15\sqrt[3]{42}$

Example 4: $2\sqrt[3]{4} \cdot 3\sqrt[3]{6}$

$= (2 \cdot 3)\sqrt[3]{24}$

$= 6\sqrt[3]{8.3} = 6\sqrt[3]{2^3 \cdot 3}$

$= 6 \cdot 2\sqrt[3]{3} = 12\sqrt[3]{3}$

Multiplying Radicals With Different Indices

To multiply radicals with different indices such as ($\sqrt{3} \cdot \sqrt[3]{3^2}$) find the least common number between the indices. This means a number that is divisible by all the index numbers.

In this example the least common number between the indices is $(2 \cdot 3) = 6$

Rewrite the radicals showing the least common number as the new index for all the radicals. To keep the value of the radical the same as it was before making this change, divide the least common number by the original index of each radical, then raise the radicand to the quotient of this division.

In this example the radical $\sqrt{3}$ becomes $\sqrt[6]{3^3}$ and the radical $\sqrt[3]{3^2}$ becomes $\sqrt[6]{3^4}$.

The next step is to multiply the new radicals and try to simplify.

$(\sqrt{3} \cdot \sqrt[3]{3^2}) = (\sqrt[6]{3^3} \cdot \sqrt[6]{3^4})$

Using the product rule, $(\sqrt[6]{3^3} \cdot \sqrt[6]{3^4}) = \sqrt[6]{3^7}$

$= \sqrt[6]{3^6 \cdot 3} = 3\sqrt[6]{3}$

80

Dividing Radicals

Quotient Rule

When the indices of radical expressions are the same, radical expressions can be divided using the quotient rule:

$$\frac{\sqrt[n]{y}}{\sqrt[n]{z}} = \sqrt[n]{\frac{y}{z}}$$

Example 1: $\dfrac{\sqrt{20}}{\sqrt{5}} = \sqrt{\dfrac{20}{5}} = \sqrt{4} = 2$

Example 2: $\dfrac{\sqrt[3]{9}}{\sqrt[3]{3}} = \sqrt[3]{\dfrac{9}{3}} = \sqrt[3]{3}$

Example 3: $\dfrac{12\sqrt{18}}{4\sqrt{6}} = 3\sqrt{\dfrac{18}{6}} = 3\sqrt{3}$

Example 4: $\dfrac{\sqrt{y^4 z^2}}{\sqrt{y^2 z}}$

$$\frac{\sqrt{y^4 z^2}}{\sqrt{y^2 z}} = \sqrt{y^2 z} = y\sqrt{z}$$

To divide a radical expression with more than one term by a radical expression that includes only one term, divide each term of the numerator by the denominator, and simplify.

Example 1: $\dfrac{\sqrt{8} + \sqrt{32} + \sqrt{12}}{\sqrt{2}}$

$$= \dfrac{\sqrt{8}}{\sqrt{2}} + \dfrac{\sqrt{32}}{\sqrt{2}} + \dfrac{\sqrt{12}}{\sqrt{2}}$$

$$= \sqrt{\dfrac{8}{2}} + \sqrt{\dfrac{32}{2}} + \sqrt{\dfrac{12}{2}}$$

$$= \sqrt{4} + \sqrt{16} + \sqrt{6}$$

$$= 2 + 4 + \sqrt{6} = 6 + \sqrt{6}$$

Example 2: $\dfrac{\sqrt{12} + \sqrt{27} + \sqrt{48}}{\sqrt{3}}$

$$= \dfrac{\sqrt{12}}{\sqrt{3}} + \dfrac{\sqrt{27}}{\sqrt{3}} + \dfrac{\sqrt{48}}{\sqrt{3}}$$

$$= \sqrt{\dfrac{12}{3}} + \sqrt{\dfrac{27}{3}} + \sqrt{\dfrac{48}{3}}$$

$$= \sqrt{4} + \sqrt{9} + \sqrt{16}$$

$$= 2 + 3 + 4 = 9$$

Example 3: $\dfrac{\sqrt{x^3} + \sqrt{x^5} + \sqrt{x^7}}{\sqrt{x}}$

$$= \sqrt{\dfrac{x^3}{x}} + \sqrt{\dfrac{x^5}{x}} + \sqrt{\dfrac{x^7}{x}}$$

$$= \sqrt{x^2} + \sqrt{x^4} + \sqrt{x^6}$$

$$= x + x^2 + x^3$$

Example 4:　$(\sqrt{6xy} + \sqrt{4yz} + \sqrt{2xz})/\sqrt{2xyz}$

$$= \frac{\sqrt{6xy}}{\sqrt{2xyz}} + \frac{\sqrt{4yz}}{\sqrt{2xyz}} + \frac{\sqrt{2xz}}{\sqrt{2xyz}}$$

$$= \sqrt{\frac{6xy}{2xyz}} + \sqrt{\frac{4yz}{2xyz}} + \sqrt{\frac{2xz}{2xyz}}$$

$$= \sqrt{\frac{3}{z}} + \sqrt{\frac{2}{x}} + \sqrt{\frac{1}{y}}$$

To divide a radical fraction by a radical fraction, such as $\dfrac{\sqrt{3}}{\sqrt{2}} \div \dfrac{\sqrt{3}}{\sqrt{5}}$

Leave the first fraction as it is, then change the divided by sign (÷) to times sign (x), reverse the numerator and the denominator of the second radical fraction, and simplify.

This means $\dfrac{\sqrt{3}}{\sqrt{2}} \div \dfrac{\sqrt{3}}{\sqrt{5}} = \dfrac{\sqrt{3}}{\sqrt{2}} \ \text{x} \ \dfrac{\sqrt{5}}{\sqrt{3}} = \dfrac{\sqrt{5}}{\sqrt{2}}$

Example 1:　$\dfrac{\sqrt{6}}{\sqrt{3}} \div \dfrac{\sqrt{24}}{\sqrt{3}}$

$$\frac{\sqrt{6}}{\sqrt{3}} \div \frac{\sqrt{24}}{\sqrt{3}} = \frac{\sqrt{6}}{\sqrt{3}} \ \text{x} \ \frac{\sqrt{3}}{\sqrt{24}}$$

$$= \frac{\sqrt{6}}{\sqrt{24}} = \sqrt{\frac{1}{4}} = \frac{1}{2}$$

Example 2: $\dfrac{\sqrt{12}}{\sqrt{5}} \div \dfrac{\sqrt{6}}{2\sqrt{5}}$

$$\dfrac{\sqrt{12}}{\sqrt{5}} \div \dfrac{\sqrt{6}}{2\sqrt{5}} = \dfrac{\sqrt{12}}{\sqrt{5}} \times \dfrac{2\sqrt{5}}{\sqrt{6}}$$

$$= \dfrac{2\sqrt{12}}{\sqrt{6}} = 2\sqrt{\dfrac{12}{6}} = 2\sqrt{2}$$

Example 3: $\dfrac{\sqrt{x^3}}{\sqrt{y}} \div \dfrac{\sqrt{x}}{\sqrt{y}}$

$$\dfrac{\sqrt{x^3}}{\sqrt{y}} \div \dfrac{\sqrt{x}}{\sqrt{y}} = \dfrac{\sqrt{x^3}}{\sqrt{y}} \times \dfrac{\sqrt{y}}{\sqrt{x}}$$

$$= \dfrac{\sqrt{x^3}}{\sqrt{x}} = \sqrt{\dfrac{x^3}{x}} = \sqrt{x^2} = x$$

Example 4: $\dfrac{\sqrt{8x^3}}{\sqrt{3y}} \div \dfrac{\sqrt{2x}}{\sqrt{27y^3}}$

$$\dfrac{\sqrt{8x^3}}{\sqrt{3y}} \div \dfrac{\sqrt{2x}}{\sqrt{27y^3}} = \dfrac{\sqrt{8x^3}}{\sqrt{3y}} \times \dfrac{\sqrt{27y^3}}{\sqrt{2x}}$$

$$= \dfrac{\sqrt{8x^3}}{\sqrt{2x}} \times \dfrac{\sqrt{27y^3}}{\sqrt{3y}} = \sqrt{\dfrac{8x^3}{2x}} \times \sqrt{\dfrac{27y^3}{3y}}$$

$$= \sqrt{4x^2} \times \sqrt{9y^2} = 2x \cdot 3y$$

84

Rationalizing the Denominator

In the cases when a radical is divided by another radical, to simplify the fraction eliminate the radical in the denominator by multiplying both the numerator and the denominator by the rationalizing factor of the denominator. The rationalizing factor is defined as the factor that changes the denominator from a radical form to a rational number.

Example 1: Rationalize the denominator of the fraction $\dfrac{4}{\sqrt{2}}$

$$\frac{4}{\sqrt{2}} = \frac{4}{\sqrt{2}} \cdot \frac{\sqrt{2}}{\sqrt{2}} = \frac{4\sqrt{2}}{2} = 2\sqrt{2}$$

Example 2: Rationalize the denominator of the fraction $\dfrac{9}{\sqrt{3y}}$

$$\frac{9}{\sqrt{3y}} = \frac{9}{\sqrt{3y}} \cdot \frac{\sqrt{3y}}{\sqrt{3y}} = \frac{9\sqrt{3y}}{3y} = \frac{3\sqrt{3y}}{y}$$

Example 3: Rationalize the denominator of the fraction $\dfrac{15}{\sqrt[3]{3}}$

$$\frac{15}{\sqrt[3]{3}} = \frac{15}{\sqrt[3]{3}} \cdot \frac{\sqrt[3]{3^2}}{\sqrt[3]{3^2}} = \frac{15\sqrt[3]{3^2}}{3} = 5\sqrt[3]{9}$$

Example 4: Rationalize the denominator of the fraction $\dfrac{\sqrt{10x^2 y^2}}{\sqrt{5xy}}$

$$\frac{\sqrt{10x^3 y^3}}{\sqrt{5xy}} = \frac{\sqrt{10x^3 y^3}}{\sqrt{5xy}} \cdot \frac{\sqrt{5xy}}{\sqrt{5xy}}$$

$$= \frac{\sqrt{50x^4 y^4}}{5xy} = \frac{5\sqrt{2x^4 y^4}}{5xy} = \frac{5x^2 y^2 \sqrt{2}}{5xy}$$

$$= xy\sqrt{2}$$

CONJUGATES

Radical expressions such as $(\sqrt{x} + \sqrt{y})$ and $(\sqrt{x} - \sqrt{y})$ when multiplied the resulting product is $(x - y)$. The term $(\sqrt{x} + \sqrt{y})$ is called the conjugate of the term $(\sqrt{x} - \sqrt{y})$, and vice versa.

To rationalize the denominator of a radical fraction in the form of $\dfrac{\sqrt{x} + \sqrt{y}}{\sqrt{b} + \sqrt{c}}$ multiply the numerator and denominator by the conjugate term of the denominator, which in this example would be $(\sqrt{b} - \sqrt{c})$.

Example 1: Divide the following radical expression and simplify:

$$\frac{1}{3 - \sqrt{5}}$$

$$\frac{1}{3 - \sqrt{5}} = \frac{1}{3 - \sqrt{5}} \cdot \frac{3 + \sqrt{5}}{3 + \sqrt{5}} = \frac{3 + \sqrt{5}}{9 - 5}$$

$$= \frac{3 + \sqrt{5}}{4}$$

Example 2: Divide the following radical expression and simplify:

$$\frac{\sqrt{2}}{3 + \sqrt{2}}$$

$$\frac{\sqrt{2}}{3 + \sqrt{2}} = \frac{\sqrt{2}}{3 + \sqrt{2}} \cdot \frac{3 - \sqrt{2}}{3 - \sqrt{2}} = \frac{3\sqrt{2} - 2}{9 - 2}$$

$$= \frac{3\sqrt{2} - 2}{7}$$

Example 3: Divide the following radical expression and simplify:

$$\frac{2+\sqrt{6}}{2-\sqrt{6}}$$

$$\frac{2+\sqrt{6}}{2-\sqrt{6}} = \frac{2+\sqrt{6}}{2-\sqrt{6}} \cdot \frac{2+\sqrt{6}}{2+\sqrt{6}}$$

$$= \frac{4+4\sqrt{6}+6}{4-6}$$

$$= \frac{10+4\sqrt{6}}{-2} = -5+4\sqrt{6}$$

Example 4: Divide the following radical expression and simplify:

$$\frac{\sqrt{5}-\sqrt{2}}{\sqrt{5}+\sqrt{2}}$$

$$\frac{\sqrt{5}-\sqrt{2}}{\sqrt{5}+\sqrt{2}} = \frac{\sqrt{5}-\sqrt{2}}{\sqrt{5}+\sqrt{2}} \cdot \frac{\sqrt{5}-\sqrt{2}}{\sqrt{5}-\sqrt{2}}$$

$$= \frac{5-2-2\sqrt{10}}{5-2}$$

$$= \frac{3-2\sqrt{10}}{3} = 1 - \frac{2\sqrt{10}}{3}$$

Exercise 6.1

Simplify and find the value of the following radicals:

1- $\sqrt{49}$ 2- $\sqrt{144}$

3- $\sqrt{-81}$ 4- $\sqrt[3]{27}$

5- $\sqrt[3]{125}$ 6- $\sqrt[3]{-64}$

7- $\sqrt[4]{x^8}$ 8- $\sqrt[4]{y^{12}}$

9- $\sqrt{\dfrac{1}{36}}$ 10- $\sqrt{\dfrac{25}{64}}$

11- $\sqrt{0.36}$ 12- $\sqrt{0.0049}$

13- $\sqrt[3]{\dfrac{1}{64}}$ 14- $\sqrt[3]{\dfrac{27}{125}}$

15- $\sqrt[3]{\dfrac{216}{1000}}$ 16- $\sqrt[3]{0.008}$

17- $\sqrt[3]{\dfrac{-1}{1000}}$ 18- $\sqrt[3]{-0.512}$

19- $\sqrt[4]{125}$ 20- $\sqrt{(y+3)^2}$

21- $\sqrt{(x-4)^2}$ 22- $\sqrt{16x^4}$

23- $\sqrt[3]{(z+6)^3}$ 24- $\sqrt[3]{x^6 y^9}$

25- $2\sqrt{45}$ 26- $\sqrt{x^5}$

27- $\sqrt{25y^4 z^8}$ 28- $z\sqrt[3]{z^3 y^6}$

29- $\sqrt[4]{y^6}$

30- $\sqrt[5]{a^6 b}$

31- $\sqrt[3]{8a^6 b^4}$

32- $\sqrt[4]{y^9 z^2}$

33- $\sqrt{64a^3 b^8}$

34- $\sqrt{z^3 (z-5)^2}$

Exercise 6.2:

Add the following radicals:

1- $5\sqrt{5} + 5\sqrt{5}$

2- $2\sqrt{6} + 4\sqrt{6}$

3- $\sqrt{3} + 8\sqrt{3}$

4- $7\sqrt{7} + \sqrt{7}$

5- $\sqrt{27} + 3\sqrt{48}$

6- $\sqrt{125} + 2\sqrt{80}$

7- $5\sqrt{12} + 2\sqrt{75}$

8- $2\sqrt{32} + 3\sqrt{50}$

9- $4\sqrt{63} + 4\sqrt{112}$

10- $6\sqrt{108} + \sqrt{147}$

11- $\sqrt{75} + \sqrt{48} + \sqrt{27}$

12- $\sqrt{125} + \sqrt{80} + \sqrt{45}$

13- $\sqrt{18} + \sqrt{27} + \sqrt{50}$

14- $2\sqrt{8} + \sqrt{32} + \sqrt{12}$

15- $\sqrt{98} + 2\sqrt{50} + \sqrt{72}$

16- $\sqrt{48} + \sqrt{75} + 2\sqrt{108}$

17- $5\sqrt[3]{125} + 2\sqrt[3]{8}$

18- $3\sqrt[3]{64} + \sqrt[3]{1}$

19- $\sqrt[3]{27} + \sqrt[3]{216}$

20- $\sqrt[3]{1000} + \sqrt[3]{729}$

21- $\sqrt[3]{16} + \sqrt[3]{54} + \sqrt[3]{128}$

22- $\sqrt[3]{192} + \sqrt[3]{81} + \sqrt[3]{24}$

23- $\sqrt[3]{24} + \sqrt[3]{16} + \sqrt[3]{54}$

24- $\sqrt[3]{40} + \sqrt[3]{24} + \sqrt[3]{135}$

25- $\sqrt[3]{48} + 2\sqrt[3]{162} + \sqrt[3]{40}$

26- $\sqrt[3]{81} + \sqrt[3]{128} + 2\sqrt[3]{250}$

27- $\sqrt{320} + \sqrt{45} + \sqrt{125}$

28- $\sqrt[3]{88} + \sqrt[3]{297} + \sqrt[3]{11}$

Exercise 6.3:

Subtract the following radicals:

1- $\sqrt{18} - \sqrt{8}$ 2- $\sqrt{9} - \sqrt{1}$

3- $\sqrt{25} - \sqrt{16}$ 4- $\sqrt{32} - \sqrt{18}$

5- $\sqrt{48} - \sqrt{64}$ 6- $\sqrt{49} - \sqrt{36}$

7- $\sqrt{96} - \sqrt{54}$ 8- $2\sqrt{81} - \sqrt{81}$

9- $6\sqrt{75} - 10\sqrt{27}$ 10- $\sqrt{63} - \sqrt{28}$

11- $\sqrt{72} - \sqrt{50} - \sqrt{32}$ 12- $\sqrt{75} - \sqrt{48} - \sqrt{8}$

13- $\sqrt{100} - \sqrt{81} - \sqrt{25}$ 14- $\sqrt{144} - \sqrt{121} - \sqrt{49}$

15- $\sqrt{147} - \sqrt{108} - \sqrt{12}$ 16- $\sqrt{252} - \sqrt{112} - \sqrt{28}$

17- $3\sqrt[3]{27} - 2\sqrt[3]{8}$ 18- $2\sqrt[3]{128} - \sqrt[3]{72}$

19- $4\sqrt[3]{125} - \sqrt[3]{64}$ 20- $\sqrt[3]{216} - 2\sqrt[3]{64}$

21- $3\sqrt[3]{8} - \sqrt[3]{8}$ 22- $2\sqrt[3]{125} - \sqrt[3]{1}$

23- $9\sqrt[3]{27} - 2\sqrt[3]{64} - 5\sqrt[3]{8}$ 24- $\sqrt[3]{128} - \sqrt[3]{54} - \sqrt[3]{16}$

25- $\sqrt[3]{192} - \sqrt[3]{24} - \sqrt[3]{81}$ 26- $\sqrt[3]{189} - \sqrt[3]{56} - \sqrt[3]{7}$

27- $\sqrt[3]{256} - \sqrt[3]{108} - \sqrt[3]{32}$ 28- $\sqrt[3]{320} - \sqrt[3]{135} - \sqrt[3]{40}$

Exercise 6.4:

Multiply the following radicals:

1- $\sqrt{5}.\sqrt{3}$ 2- $\sqrt{2}.\sqrt{27}$

3- $\sqrt{20}.\sqrt{5}$ 4- $\sqrt{7}.\sqrt{7}$

5- $\sqrt{14}.\sqrt{7}$ 6- $\sqrt{6}.\sqrt{12}$

7- $\sqrt{18}.\sqrt{8}$ 8- $\sqrt{8}.\sqrt{6}$

9- $3\sqrt{3}.\sqrt{15}$ 10- $6\sqrt{2}.(-\sqrt{32})$

11- $\sqrt{2y}.\sqrt{2y}$ 12- $\sqrt{5xy}.\sqrt{5x^2 y^3}$

13- $\sqrt{8z}.\sqrt{8zy^2}$ 14- $\sqrt{12x^4 y}.\sqrt{6y^3}$

15- $\sqrt{12x^3 y}.\sqrt{12xy^3}$ 16- $\sqrt{3}.\sqrt[3]{3^2}$

17- $\sqrt[3]{x^4}.\sqrt[6]{x^4}$ 18- $\sqrt[4]{7^3}.\sqrt[4]{7^5}$

19- $\sqrt[3]{5^2}.\sqrt[4]{5^3}$ 20- $3\sqrt[3]{3}.\sqrt[3]{6}$

21- $2\sqrt[3]{5}.(-3\sqrt[3]{25})$ 22- $5\sqrt[3]{4}.(2\sqrt[3]{16})$

23- $\sqrt{y}.\sqrt[4]{y^3}$ 24- $\sqrt{yz}.\sqrt[3]{y^2}$

25- $\sqrt{(y-3)}.\sqrt{(y+3)}$ 26- $\sqrt[3]{-72}.\sqrt[3]{-3}$

27- $\sqrt[4]{z^3}.\sqrt[4]{z^5}$ 28- $\sqrt[4]{8y^3 z}.\sqrt[4]{2yz^3}$

29- $(2\sqrt{2}+3\sqrt{5})(2\sqrt{2}-3\sqrt{5})$ 30- $(\sqrt{15}-\sqrt{5})(\sqrt{15}+\sqrt{5})$

Exercise 6.5:

Divide the following radicals and simplify:

1- $\dfrac{\sqrt{8}}{\sqrt{2}}$ 2- $\dfrac{\sqrt{12}}{\sqrt{3}}$

3- $\dfrac{\sqrt{27}}{\sqrt{3}}$

4- $\dfrac{\sqrt{80}}{\sqrt{5}}$

5- $\dfrac{\sqrt{12}}{\sqrt{6}}$

6- $\dfrac{\sqrt{21}}{\sqrt{7}}$

7- $\dfrac{\sqrt{15}}{\sqrt{3}}$

8- $\dfrac{\sqrt{125}}{\sqrt{5}}$

9- $\dfrac{\sqrt{500}}{\sqrt{5}}$

10- $\dfrac{\sqrt[3]{48}}{\sqrt[3]{6}}$

11- $\dfrac{\sqrt[3]{54}}{\sqrt[3]{2}}$

12- $\dfrac{\sqrt[3]{9}}{\sqrt[3]{3}}$

13- $\dfrac{\sqrt[3]{-81}}{\sqrt[3]{3}}$

14- $\dfrac{\sqrt[3]{-75}}{\sqrt[3]{5}}$

15- $\dfrac{\sqrt{5}}{\sqrt{45}}$

16- $\dfrac{\sqrt{x^3 y^3}}{\sqrt{xy}}$

17- $\dfrac{\sqrt{a^5 b^2}}{\sqrt{ab^2}}$

18- $\dfrac{\sqrt[3]{y^5 z^4}}{\sqrt[3]{y^2 z}}$

19- $\dfrac{\sqrt{12y^3}}{\sqrt{3}} \div \dfrac{\sqrt{3y}}{\sqrt{3}}$

20- $\dfrac{\sqrt{28}}{\sqrt{2}} \div \dfrac{\sqrt{7}}{\sqrt{2}}$

21- $\dfrac{\sqrt{y^3}}{\sqrt{y-3}} \div \dfrac{\sqrt{y}}{\sqrt{y-3}}$

22- $\dfrac{\sqrt{3x^2}}{\sqrt{2y}} \div \dfrac{\sqrt{12x^2}}{\sqrt{4y^3}}$

23- $\dfrac{\sqrt{8}+\sqrt{18}}{\sqrt{2}}$

24- $\dfrac{\sqrt{12}+\sqrt{48}}{\sqrt{3}}$

25- $\dfrac{\sqrt{20} - \sqrt{35} + \sqrt{45}}{\sqrt{5}}$

26- $\dfrac{\sqrt[3]{16} + \sqrt[3]{54}}{\sqrt[3]{2}}$

27- $\dfrac{\sqrt[3]{81} + \sqrt[3]{24} + \sqrt[3]{192}}{\sqrt[3]{3}}$

28- $\dfrac{\sqrt[3]{108} - \sqrt[3]{32} + \sqrt[3]{256}}{\sqrt[3]{4}}$

29- $\dfrac{\sqrt{x^3 y^3} + \sqrt{x^5 y^5}}{\sqrt{xy}}$

30- $\dfrac{\sqrt{x^4 y^3} - \sqrt{x^6 y^5}}{\sqrt{x^2 y}}$

31- $\dfrac{\sqrt[3]{x^6 y^5} + \sqrt[3]{x^{11} y^{10}}}{\sqrt[3]{x^2 y^2}}$

32- $\dfrac{2\sqrt{a^3 b^6} + 3\sqrt{ab^2} - \sqrt{a^5 b^6}}{\sqrt{ab^2}}$

Exercise 6.6:

Rationalize the following radicals and simplify:

1- $\dfrac{3}{\sqrt{5}}$

2- $\dfrac{4}{\sqrt{2}}$

3- $\dfrac{8}{\sqrt{8}}$

4- $\dfrac{6}{\sqrt{3}}$

5- $\dfrac{2}{\sqrt[3]{3}}$

6- $\dfrac{5}{\sqrt[3]{4}}$

7- $\dfrac{1}{\sqrt[3]{6}}$

8- $\dfrac{10}{\sqrt[3]{5}}$

9- $\dfrac{8}{\sqrt[4]{4}}$

10- $\dfrac{2}{\sqrt{2y}}$

11- $\dfrac{3}{\sqrt{3z}}$

12- $\dfrac{x}{\sqrt{x^2 y}}$

13- $\dfrac{xy}{\sqrt{2xy}}$

14- $\dfrac{4yz}{\sqrt{2y^2 z}}$

15- $\dfrac{6x^2 y}{\sqrt{2x^2 y}}$

16- $\dfrac{3y^3 z}{\sqrt{y^2 z^3}}$

17- $\dfrac{1}{4-\sqrt{5}}$

18- $\dfrac{2}{3-\sqrt{3}}$

19- $\dfrac{\sqrt{5}}{1-\sqrt{5}}$

20- $\dfrac{\sqrt{6}}{2+\sqrt{6}}$

21- $\dfrac{3\sqrt{3}}{\sqrt{3}+\sqrt{2}}$

22- $\dfrac{5\sqrt{7}}{\sqrt{7}-\sqrt{6}}$

23- $\dfrac{3-\sqrt{2}}{2\sqrt{2}-\sqrt{3}}$

24- $\dfrac{\sqrt{5}-\sqrt{2}}{\sqrt{5}+2\sqrt{2}}$

25- $\dfrac{3+\sqrt{z}}{\sqrt{2}-\sqrt{z}}$

26- $\dfrac{\sqrt{a}-\sqrt{b}}{\sqrt{2a}-\sqrt{b}}$

27- $\dfrac{1+\sqrt{7}}{\sqrt{7}-\sqrt{2}}$

28- $\dfrac{\sqrt{3z}-\sqrt{2y}}{\sqrt{2z}+\sqrt{2y}}$

CHAPTER 7

Solving Linear Equations

Definition of Linear Equations

Linear equations are defined as equations with terms that have only one variable, and the terms have exponents of one.

The following are examples of linear equations:

Example 1: $2x = 4$

Example 2: $x + 3 = 5$

Example 3: $x - 3x = 10$

Example 4: $2x - 4x = 12$

Example 5: $3x + 5y = 11$

Example 6: $5x - 3y + 2z = 15$

The following are examples of non-linear equations:

Example 1: $a + 5bc = 6$

Note that the second term has two variables

Example 2: $2y^2 - 7y + 5 = 0$

Note that the first term has an exponent of 2

Example 3: $4abc + 3bc = 9$

Note that the first term has three variables, and the second term has two variables.

Definition of Equivalent Equations

The word "Equivalent" here means equal in value. This means that equivalent equations are equations that have the same solution value.

The following are examples of equivalent equations:

Example 1: $5x = 15$, and $x = 3$

                The two equations have the same solution value of 3.

Example 2: $4x + 5 = 13$, and $x = 2$

                The two equations have the same solution value of 2.

Example 3: $2y - y = 9$, and $y = 9$

                The two equations have the same solution value of 9.

Example 4: $3y + 2y - y = 20$, and $y = 5$

                The two equations have the same solution value of 5.

Equivalent equations are very helpful in solving linear equations, and can be created using the following methods:

1-   Adding the same quantity or variable to each side of the equation. This is called the addition property.

2-   Subtracting same quantity or variable from each side of the equation. This is called the subtraction property.

3-   Multiplying each side of the equation by the same quantity. This is called the multiplication property.

        However, when each side of the equation is multiplied by a variable, or raised to a power bigger than 1, the new equation may not be equivalent to the original equation. This will be explained in more details later on.

4-   Dividing each side of the equation by the same quantity or variable. This is called the division property.

Solving Linear Equations by combining similar terms

If the linear equation includes the same variable in several terms on one side of the equation, the linear equation can be easily solved by combining similar terms

Example 1:   Find the value of y in the following equation:

$$2y + 5y - y = 6$$

Combine similar terms on the left side of the equation. Adding the first two terms results in 7y

Subtracting the third term makes the net value of the left side of the equation equals 6y.

Therefore, this linear equation can be simplified to read

$6y = 6$, and the solution will be y =1

Example 2:   Find the value of x in the following equation:

$$12x - 2x - 7x = 9$$

Combine similar terms on the left side of the equation. Subtracting the first two terms results in $10x$

Subtracting the third term makes the net value of the left side of the equation equals $3x$.

Therefore, this linear equation can be simplified to read

$3x = 9$, and the solution will be $x = 3$

Solving Linear Equations by the Addition Property

If one or both sides of the linear equation include terms that include a variable and numbers, the equation can be solved by adding a variable or a number to each side of the equation so that the equation is simplified by having only the variable on one side of the equation and only the number on the other side of the equation.

Example 1:   Find the value of x in the following equation:

$$x - 1 = 9$$

To eliminate the number on the left side of the equation add (+ 1) to both sides of the equation.

$$x - 1 + 1 = 9 + 1$$

$$x = 10$$

Example 2:   Find the value of y in the following equation:

$$5 + y = 8$$

To eliminate the number on the left side of the equation add (− 5) to both sides of the equation.

$$5 + y - 5 = 8 - 5$$

$$y = 3$$

Example 3:   the value of b in the following equation:

$$9\,b + 7 = 2\,b$$

Note that each side of the equation includes a term of (b).

To find the value of (b), first, let us eliminate the term on the right side of the equation, which in fact is a way to combine the similar terms.

To eliminate the term (2b) on the right side of the equation add (− 2b) to both sides of the equation.

$$9\,b + 7 - 2b = 2\,b - 2b$$

$$7\,b + 7 = 0$$

Second, let us eliminate the number on the left side of the equation by adding (− 7) to both sides of the equation.

$$7\,b + 7 - 7 = 0 - 7$$

$$7\,b = -7$$

Divide each side of the equation by 7

$$b = -1$$

Solving Linear Equations by the Multiplication Property

If one or both sides of the linear equation include terms that include whole numbers or fractions, the equation can be solved by multiplying each side of the equation by the inverse of the number so that the equation is simplified by having only the variable on one side of the equation and only the number on the other side of the equation.

Example 1: Find the value of x in the following equation:

$3x = 15$

Multiply each side of the equation by the inverse of the coefficient of x, which is 1/3

$1/3 . (3\ x) = 1/3 . (15)$

$x = 5$

Example 2: Find the value of y in the following equation:

$2y + 4y = 12$

Combine similar terms on the left side of the equation

$(2y + 4y) = 6y$

$6y = 12$

Multiply each side of the equation by the inverse of the coefficient of y, which is 1/6

$1/6 . (6y) = 1/6 . (12)$

$y = 2$

Example 3: Find the value of z in the following equation:

$z / 7 = 4$

Multiply each side of the equation by the inverse of the coefficient of z, which is 7

$$(7) \cdot (z \, / \, 7) = (7) \cdot (4)$$

$$z = 28$$

When each side of the equation is multiplied by a variable, or raised to a power bigger than 1, the new equation may not be equivalent to the original equation.

Example 1: $4x = 8$

If each side of the equation is multiplied by x, the equation becomes $4x \cdot x = 8x$ or $4x^2 = 8x$

To solve the first equation, multiply each side of the equation by (1/4)

$(1/4) \cdot 4x = (1/4) \cdot 8$

$x = 2$

However, the new equation obtained by multiplying each side of the equation by x, which is ($4x^2 = 8x$) has two solutions, 0 and 2. Therefore, the new equation is not equivalent to the first equation.

Example 2: $y = 9$

If each side of the equation is multiplied by y, the equation becomes $y \cdot y = 9y$ or $y^2 = 9y$

The new equation obtained by multiplying each side of the equation by y, which is ($y^2 = 9y$) has two solutions, 0 and 9. Therefore, the new equation is not equivalent to the first equation.

Solving Linear Equations by the Division Property

If one or both sides of the linear equation include terms that include whole numbers or fractions, the equation can be solved by dividing each side of the equation by the number so that the equation is simplified by having only the variable on one side of the equation and only the number on the other side of the equation.

Example 1:   Find the value of x in the following equation:

$6x = 12$

Divide each side of the equation by 6

$6x / 6 = 12 / 6$

$x = 2$

Example 2: $5y + 6y = 33$

Combine the similar terms on the left side of the equation.

$(5y + 6y) = 33$

$11y = 33$

Divide each side of the equation by 11

$11y / 11 = 33 / 11$

$y = 3$

Solving Linear Equations that include decimals or fractions

In cases when the coefficient of the variable is in a decimal format such as (2.4 y), to solve the equation and find the value of the variable, change the decimal to a regular fraction.

Example 1: $2.4 y = 12$

Convert the coefficient of y to a fraction

$$2.4 = \frac{24}{10}$$

The equation becomes $\dfrac{24}{10} y = 12$

Multiply each side of the equation by the inverse of the coefficient of y.

The inverse of $\dfrac{24}{10}$ is $\dfrac{10}{24}$

$$\dfrac{10}{24} \cdot \dfrac{24}{10} y = \dfrac{10}{24} \cdot 12$$

$$y = \dfrac{10}{2}$$

$$y = 5$$

Example 2: $0.7 x = 2.8$

Convert the coefficient of x to a fraction

$$0.7 = \dfrac{7}{10}$$

To simplify the right side of the equation, convert the decimal to a fraction.

$$2.8 = \dfrac{28}{10}$$

The equation becomes $\dfrac{7}{10} x = \dfrac{28}{10}$

Multiply each side of the equation by the inverse of the coefficient of x.

The inverse of $\dfrac{7}{10}$ is $\dfrac{10}{7}$

$$\dfrac{10}{7} \cdot \dfrac{7}{10} x = \dfrac{10}{7} \cdot \dfrac{28}{10}$$

$$x = \dfrac{28}{7} = 4$$

Example 3: Solve the equation $\dfrac{3x}{x+2} = 5$

Multiply each side of the equation by $(x+2)$

$3x = 5(x+2)$

$3x = 5x + 10$

Add $(-5x)$ to each side of the equation

$3x - 5x = 5x - 5x + 10$

$-2x = 10$

Divide each side of the equation by (-2)

$x = -5$

Example 4: Solve the equation $\dfrac{9}{x} - \dfrac{2}{3x} - \dfrac{1}{3x} = 8$

Multiply each side of the equation by $3x$

$27 - 2 - 1 = 24x$

$24 = 24x$

$x = 1$

Example 5: Solve the equation $\dfrac{x-2}{2x+1} = 3$

Multiply each side of the equation by $(2x+1)$

$x - 2 = 3(2x+1)$

$x - 2 = 6x + 3$

Add $-6x$ to each side of the equation

$x - 2 - 6x = 6x - 6x + 3$

$-5x - 2 = 3$

Add 2 to each side of the equation

$$-5x - 2 + 2 = 3 + 2$$

$$-5x = 5$$

Divide each side of the equation by -5

$$x = -1$$

In cases when the coefficient of the variable is a fraction such as ($\frac{3}{5}$), to solve the equation and find the value of the variable, multiply each side of the equation by the inverse of the variable coefficient, which would be ($\frac{5}{3}$) for this example.

Example 1: $\qquad \frac{3}{5}x = 6$

Multiply each side of the equation by the inverse of the coefficient of y.

The inverse of $\frac{3}{5}$ is $\frac{5}{3}$

$$\frac{5}{3} \cdot \frac{3}{5}x = \frac{5}{3} \cdot 6$$

$$x = \frac{30}{3}$$

$$x = 10$$

Example 2: $\qquad \frac{11}{2}y = \frac{33}{8}$

Multiply each side of the equation by the inverse of the coefficient of y.

The inverse of $\frac{11}{2}$ is $\frac{2}{11}$

$$\frac{2}{11} \cdot \frac{11}{2} \, y = \frac{2}{11} \cdot \frac{33}{8}$$

$$y = \frac{3}{4}$$

Solving Linear Equations that have variables on both sides

In cases when both sides of a linear equation include a variable, to solve the equation and find the value of the variable, form an equivalent equation that has the variable on one side of the equation, this can be accomplished by adding the additive inverse of the other term (other than the variable) to each side of the equation.

Example 1: $3x - 6 = x$

To eliminate the term (-6) from the left side of the equation, add the additive inverse of (-6), which is (6) to each side of the equation.

$3x - 6 + 6 = x + 6$

$3x = x + 6$

To eliminate the term (x) from the right side of the equation, add $(-x)$ to each side of the equation.

$3x - x = x + 6 - x$

$2x = 6$

Divide each side of the equation by 2

$x = 3$

Example 2: $3y + 22 = 8y + 2$

To eliminate the term (22) from the left side of the equation, add the additive inverse of (22), which is (-22) to each side of the equation.

$$3y + 22 - 22 = 8y + 2 - 22$$

$$3y = 8y - 20$$

To eliminate the term (8 y) from the right side of the equation, add (− 8 y) to each side of the equation.

$$3y - 8y = 8y - 20 - 8y$$

$$-5y = -20$$

$$y = \frac{-20}{-5}$$

$$y = 4$$

Solving Linear Equations that include radicals

To simplify and solve an equation that includes radicals, eliminate the radicals by raising both sides of the equation to the power of the radical index. If after completing this step, the equation still includes radicals, repeat the aforementioned step to eliminate the remaining radicals.

Example 1: Find the value of x in the following equation:

$$\sqrt{3x - 2} = 5$$

Square both sides of the equation

$$(\sqrt{3x - 2})^2 = 5^2$$

$$3x - 2 = 25$$

$$3x = 27$$

Divide each side of the equation by 3

$$x = 9$$

Example 2: Find the value of y in the following equation:

$$\sqrt{y-4} + \sqrt{y} = 2$$

Add $(-\sqrt{y})$ to each side of the equation

$$\sqrt{y-4} + \sqrt{y} - \sqrt{y} = 2 - \sqrt{y}$$

$$\sqrt{y-4} = 2 - \sqrt{y}$$

Square both sides of the equation

$$(\sqrt{y-4})^2 = (2 - \sqrt{y})^2$$

$$y - 4 = 4 - 4\sqrt{y} + y$$

$$4\sqrt{y} = 8$$

Divide each side of the equation by 4

$$\sqrt{y} = 2$$

Square both sides of the equation

$$y = 4$$

Solving Linear Equations that include absolute values

The absolute value of an unknown such as x, written $|x|$,
is equal to x, if x is ≥ 0
or is equal to $-x$, if x is < 0

Linear equations that include absolute values are solved using the steps shown in the examples below.

Example 1: Find the values of x in the following equation:

$$|x - 1| = 9$$

Since the value of $x - 1$ may be ≥ 0, or it may be < 0, we have to consider both of these two possibilities to find the values of x.

➢ If $x - 1 \geq 0$, then $x \geq 1$

The absolute value of $x - 1$, which is written as $\left| x - 1 \right|$, equals $x - 1$

Now we can write the original equation as:

$x - 1 = 9$

Which yields a value of $x = 10$

➢ If $x - 1 < 0$, then $x < 1$

In this case the absolute value of $x - 1$, which is written as $\left| x - 1 \right|$, equals $-(x - 1) = -x + 1$

Now we can write the original equation as:

$-x + 1 = 9$

Multiply both sides of the equation by -1

$x - 1 = -9$

$x = -8$

The two values of x are 10 and -8

Example 2: Find the values of x in the following equation:

$\left| 3x - 6 \right| = x + 8$

Since the value of $3x - 4$ may be ≥ 0, or it may be < 0, we have to consider both of these two possibilities to find the values of x.

➢ If $3x - 6 \geq 0$, then $x \geq 2$

The absolute value of $3x - 6$, which is written as $|3x - 6|$, equals $3x - 6$

Now we can write the original equation as:

$3x - 6 = x + 8$

$2x = 14$

Which yields a value of $x = 7$

➢ If $3x - 6 < 0$, then $3x < 6$ or $x < 2$

In this case the absolute value of $3x - 6$, which is written as $|3x - 6|$, equals $-(3x - 6) = -3x + 6$

Now we can write the original equation as:

$-3x + 6 = x + 8$

Multiply both sides of the equation by -1

$3x - 6 = -x - 8$

Add x to each side of the equation

$3x - 6 + x = -x - 8 + x$

$4x - 6 = -8$

Add 6 to each side of the equation

$4x - 6 + 6 = -8 + 6$

$4x = -2$

Or $x = -\dfrac{1}{2}$

The two values of x are 7 and $-\dfrac{1}{2}$

Solving Word Problems that can be expressed

In the form of Linear Equations

Word problems include information about certain relationships of unknowns. There are many types of word problems and may include, but not limited to, coin problems, even or odd numbers, geometric figure properties, distances, and percent.

The first step in solving these types of linear equations is to translate the given words and the relationships to a linear equation. The next step is to use the known methods to solve the equation and find the value of the unknown.

The following are examples of how to translate the words to an equation:

Example 1: When you add three to a number.

To form a linear equation, assume the number is x, then these words can be written as $x + 3$

Example 2: Two numbers, one of them is 4 more than the other number.

Assume the first number to be x, then the second number will be $x + 4$

Example 3: Two consecutive even numbers.

Note that consecutive even numbers such as 2, 4, 6, 8, 10, ... etc. each one of them is two more than the previous number.

So, if we assume the first number to be x, the second number would be $x + 2$

Example 4: The value of x dimes in cents.

Each dime = 10 cents

Therefore, the number of cents in x dimes = $10\ x$

Example 5: If a train travels at a speed of 60 miles per hour, how far will it be from where it started after x hours?

Note that the train travels 60 miles every hour, therefore, the distance travelled after x hours $= 60\ x$

Example 6: How many cents are in x quarters?

Each quarter equals 25 cents, therefore x quarters equal $25\ x$ cents

Example 7: If the amount of salt in a water solution is 10%, what is the amount of salt in x gallons?

Each gallon contains 10% salt, therefore, x gallons contain An amount of salt equal to $x\ .\ 10\%$

$$x\ .\ 10\% = x\ .\ \frac{10}{100} = \frac{x}{10}$$

Example 8: If John invests x dollars in a certificate of deposit that pays an interest of 5% per year. How much interest will John get after one year.

The interest amount equals $x\ .\ 5\%$

$$x\ .\ 5\% = x\ .\ \frac{5}{100}$$

Example 9: If the length of each side of a square equals x, what is the length of the perimeter of this square.

Perimeter is defined as the total length of all the sides.

Since a square has four sides, and the length of each side equals x, therefore, the length of the perimeter $= x + x + x + x = 4\ x$

Example 10:   If the width of a tennis field is 51 feet shorter than its length, how long is the field width?

Assume the length of the tennis field equals x, therefore, the width equals $x - 51$

The next step after translating the words into an equation is to solve the equation to find the value of the unknowns.

The following are examples of various types of word problems:

Example 1:   if 6 is added to two thirds of a number the answer is 10. Find the number.

Assume the number $= x$

$$6 + (\frac{2}{3} . x) = 10$$

To eliminate the 6 on the left side of the equation add (-6) to each side of the equation.

$$6 + (\frac{2}{3} . x) - 6 = 10 - 6$$

$$\frac{2}{3} . x = 4$$

To eliminate the fraction coefficient of x on the left side of the equation multiply each side of the equation by the inverse of the coefficient. The inverse of $\frac{2}{3}$ is $\frac{3}{2}$

$$\frac{3}{2} . \frac{2}{3} . x = \frac{3}{2} . 4$$

$$x = 6$$

Example 2: Two numbers, the second number is 7 more than the first number, and the sum of the two numbers is 21. Find the two numbers.

Assume the first number $= x$

The second number would be $x + 7$

The sum of the two numbers $= 21$

$(x) + (x + 7) = 21$

$x + x + 7 = 21$

$2x + 7 = 21$

To eliminate $(+ 7)$ on the left side of the equation add $(- 7)$ to each side of the equation.

$2x + 7 - 7 = 21 - 7$

$2x = 14$

Divide each side of the equation by 2

$x = 7$

So, the first number is 7

and the second number $= x + 7 = 7 + 7 = 14$

Example 3: Mira bought a television on sale for $480.00. The sale price was 20% less than the regular price. How much was the regular price of the television?

Assume the regular price of the television $= y$

Regular price $-$ (regular price x $\dfrac{20}{100}$) $=$ Sale price

$y - (y \cdot \dfrac{20}{100}) = 480$

113

$$y - \frac{y}{5} = 480$$

Multiply each side of the equation by 5

$5\,y - y = 2400$

$4\,y = 2400$

Divide each side of the equation by 4

$y = 600$

The regular price of the television was $600.00

Example 4: Yolanda invested $15,000.00 at an interest rate of 8% per year. What amount of money should Yolanda invest at 6% per year to receive a total amount of interest of $1,800.00 from both investments after one year?

Assume the amount she should invest at 6% = x

The interest amount received from the first investment

$$= 15,000 \ x \ \frac{8}{100} = \$1,200.00$$

The interest amount Yolanda should receive from the second investment = $1,800 - 1,200 = \$600.00$

$$x \cdot \frac{6}{100} = 600$$

Multiply each side of the equation by $\dfrac{100}{6}$

$$x \cdot \frac{6}{100} \cdot \frac{100}{6} = 600 \cdot \frac{100}{6}$$

$x = 10,000$

The amount of money Yolanda should invest at 6% = $10,000.

Example 5: George and John are in a bicycle race. They both start from the same point heading in the same direction. If George's speed is 14 miles per hour, and John's speed is 10 miles per hour. In how many hours will they be 7 miles apart?

The relationship between distance, speed, and time is identified by the following equation:

Distance = Speed x Time

Assume the distance traveled by John = x

The distance traveled by George = $x + 7$

Substitute John's information (distance and speed) in the distance, speed, time equation.

$x = 10 \cdot t$ (t is the time)

Substitute George's information (distance and speed) in the distance, speed, time equation.

$x + 7 = 14 \cdot t$ (t is the time)

Since $x = 10\,t$, substitute 10 t for x in the above equation.

$10\,t + 7 = 14\,t$

Add $(-10\,t)$ to each side of the equation.

$10\,t + 7 - 10\,t = 14\,t - 10\,t$

$7 = 4\,t$

Divide each side of the equation by 4

$t = \dfrac{7}{4} = 1\dfrac{3}{4}$ hours = 1 hour 45 minutes.

Example 6: Two trains are leaving Chicago station, the first train is heading east, and the second train is heading west. The speed of the second train is 20 miles per hour less than the speed of the first train. After 2.5 hours, the two trains were 300 miles apart. What was the speed of each train?

Assume the speed of the first train = S

The speed of the second train = S – 20

Distance = Speed x Time

After 2.5 hours

The distance traveled by the first train = S . 2.5 = 2.5 S

The distance traveled by the second train = (S – 20) . 2.5

= 2.5 S – 50

The sum of the two distances = 300

2.5 S + (2.5 S – 50) = 300

2.5 S + 2.5 S – 50 = 300

5 S – 50 = 300

Add 50 to each side of the equation

5 S – 50 + 50 = 300 + 50

5 S = 350

Divide each side of the equation by 5

$$S = \frac{350}{5} = 70$$

The speed of the first train = 70 miles per hour

The speed of the second train = 70 – 20

= 50 miles per hour.

Example 7:   John and George are two brothers. George's age is one half of John's age plus one. If the sum of the two brothers' ages is 19, how old are they now?

Assume John's age = y

George's age = $\dfrac{y}{2} + 1$

Since the sum of the two brothers ages is 19

$\therefore\ y + (\dfrac{y}{2} + 1) = 19$

$y + \dfrac{y}{2} + 1 = 19$

multiply each side of the equation by 2

$2\,y + y + 2 = 38$

$3y + 2 = 38$

Add – 2 to each side of the equation

$3y + 2 - 2 = 38 - 2$

$3\,y = 36$

Divide each side of the equation by 3

$y = 12$

$\therefore$ John's age = 12 years

George's age = $\dfrac{y}{2} + 1 = \dfrac{12}{2} + 1$

$= 6 + 1 = 7$ years

Example 8: Mira is 8 years younger than Caroline. 23 years ago
Caroline was twice as old as Mira. How old are they now?

Assume Caroline's age now = x

Mira's age now = $x - 8$

Caroline's age 23 years ago = $x - 23$

Mira's age 23 years ago = $(x-8) - 23 = x - 31$

23 years ago Caroline was twice as old as Mira

$\therefore \; x - 23 = 2\,[x - 31]$

$x - 23 = 2x - 62$

Add - $2x$ to each side of the equation

$x - 23 - 2x = 2x - 62 - 2x$

$-x - 23 = -62$

Add 23 to each side of the equation

$-x - 23 + 23 = -62 + 23$

$-x = -39$

$x = 39$

$\therefore$ Caroline's age now = 39 years

Mira's age now = $x - 8 = 39 - 8 = 31$ years

Example 9: Yola has $6.50 in nickels and dimes. If the total number of coins is 80, how many nickels and how many dimes does Yola have?

Assume the number of dimes = y

$\therefore$ the number of nickels = 80 – y

The value of y dimes in cents = 10 y

The value of (80 – y) nickels in cents

= 5 (80 – y) = 400 – 5 y

The total value of the coins is $6.50 = 650 cents

10 y + (400 – 5 y) = 650

10 y + 400 – 5 y = 650

5 y + 400 = 650

Add – 400 to each side of the equation

5 y + 400 – 400 = 650 – 400

5 y = 250

Divide each side of the equation by 5

y = 50

∴ the number of dimes = 50

The number of nickels = 80 – y

= 80 – 50 = 30

Example 10:   Lydia saved quarters and dimes. The number of quarters is 10 more than the number of dimes. If the total value of the coins is $9.50, how many quarters and dimes does Lydia have?

Assume the number of quarters = x

∴ the number of dimes = $x - 10$

The value of x quarters in cents = 25 x

The value of ($x - 10$) dimes in cents

= 10 ($x - 10$) = 10 $x - 100$

The total value of the coins is $9.50 = 950 cents

25 x + (10 $x - 100$) = 950

25 x + 10 $x - 100 = 950$

35 $x - 100 = 950$

Add 100 to each side of the equation

35 $x - 100 + 100 = 950 + 100$

35 x = 1050

Divide each side of the equation by 35

$x = 30$

∴ the number of quarters = 30

The number of dimes = $x - 10$

$= 30 - 10 = 20$

Example 11: The length of a bridge is 18 feet more than three times its width. The sum of the bridge length and width is 114 feet. Find the length and width of the bridge.

Assume the bridge width = y

The bridge length = $3y + 18$

The sum of the bridge length and width is 114 feet

∴ $y + (3y + 18) = 114$

$y + 3y + 18 = 114$

$4y + 18 = 114$

Add $- 18$ to each side of the equation

$4y + 18 - 18 = 114 - 18$

$4y = 96$

Divide each side of the equation by 4

$y = 24$

∴ the bridge width = 24 feet

The bridge length = $3y + 18$

$= (3 \times 24) + 18 = 72 + 18 = 90$ feet

Example 12: The width of a desk is 12 inches less than half the desk length. If the sum of the desk length and width is 132 inches, find the length and width of the desk.

Assume the desk length $= x$

The desk width $= \dfrac{x}{2} - 12$

The sum of the desk length and width is 132

$\therefore\ x + (\dfrac{x}{2} - 12) = 132$

$x + \dfrac{x}{2} - 12 = 132$

Multiply each side of the equation by 2

$2x + x - 24 = 264$

$3\ x - 24 = 264$

Add 24 to each side of the equation

$3\ x - 24 + 24 = 264 + 24$

$3\ x = 288$

Divide each side of the equation by 3

$x = 96$

$\therefore$ the desk length $= 96$ inches

The desk width $= \dfrac{x}{2} - 12$

$= \dfrac{96}{2} - 12 = 48 - 12 = 36$ inches

Example 13: The sum of two consecutive odd numbers is 112. Find the two numbers.

Assume the first odd number $= y$

Since they are consecutive odd numbers

∴ The second odd number = y + 2

The sum of two consecutive odd numbers is 112

y + (y + 2) = 112

y + y + 2 = 112

2 y + 2 = 112

Add − 2 to each side of the equation

2 y + 2 − 2 = 112 − 2

2y = 110

Divide each side of the equation by 2

y = 55

∴ the first odd number = 55

The second odd number = y + 2

= 55 + 2 = 57

Example 14: Find two consecutive even numbers such that the difference of their squares is 76.

Assume the first even number = x

The second even number = $x + 2$

The difference of their squares is 76

$(x + 2)^2 - x^2 = 76$

$x^2 + 4x + 4 - x^2 = 76$

$4x + 4 = 76$

Add − 4 to each side of the equation

$4x + 4 − 4 = 76 − 4$

$4x = 72$

Divide each side of the equation by 4

$x = 18$

$\therefore$ the first even number $= 18$

The second even number $= x + 2$

$= 18 + 2 = 20$

Example 15: How many gallons of water must be added to 20 gallons of 80% concentration orange juice to make an orange juice mixture of 50% concentration?

The original orange juice mixture is as follows:

Volume $= 20$ gallons Orange concentration $= 80\%$

The water to be added is as follows:

Volume $= y$ gallons Orange concentration $= 0\%$

The resulting mixture is as follows:

Volume $= (y + 20)$ Orange concentration $= 50\%$

The volume of the resulting mixture multiplied by its orange concentration must equal the volume of the original mixture multiplied by its orange concentration plus the volume of the water added multiplied by its orange concentration.

This leads to the following equation:

$$(y + 20) \frac{50}{100} = 20 \left(\frac{80}{100}\right) + y \times 0$$

Multiply each side of the equation by 100

$(y + 20) \, 50 = 20 \, (80)$

$50 \, y + 1000 = 1600$

Add $- 1000$ to each side of the equation

$50y + 1000 - 1000 = 1600 - 1000$

$50y = 600$

Divide each side of the equation by 50

$y = 12$

$\therefore$ the number of gallons of water to be added is 12.

Example 16:   If you mix 20 pounds of a 95% gold alloy with 8 pounds of a 25% gold alloy, what will the gold percentage be in the resulting mixture?

The first alloy is as follows:

Weight = 20 pounds   Percent of gold = 95%

The second alloy is as follows:

Weight = 8 pounds   Percent of gold = 25%

The resulting mixture is as follows:

Weight = 28 pounds   Percent of gold = x %

The weight of the resulting mixture multiplied by its percent of gold must equal the weight of the first alloy multiplied by its percent of gold plus the weight of the second alloy multiplied by its percent of gold.

This leads to the following equation:

$$28 \left(\frac{x}{100} \right) = 20 \left(\frac{95}{100} \right) + 8 \left(\frac{25}{100} \right)$$

Multiply each side of the equation by 100

$28x = 20 \times 95 + 8 \times 25$

$28x = 1900 + 200 = 2100$

Divide each side of the equation by 28

$x = 75$

$\therefore$ the percent of gold in the resulting mixture = 75%

Exercise 7.1

Combine similar terms in the following:

1-	$7x + 6x$		2-	$19a + 16a$
3-	$14y - 5y$		4-	$42z - 20z$
5-	$x + 11x$		6-	$29y + y$
7-	$-8x - 12x$		8-	$-10a - 2a$
9-	$-9y + 15y$		10-	$-11z + 4z$
11-	$x + 17 + 4x + 3$		12-	$5 + 4a + 16 + 5a$
13-	$12y - 1 + y + 4$		14-	$6z - 7 + 4z + 9$
15-	$18x - 3 - 13x + 6$		16-	$22z - 9 - 11z + 7$
17-	$17y - 7 - 2y - 3$		18-	$20a - 5 - a - 4$
19-	$-x - 13 - 9x - 12$		20-	$-3z - 16 - 7z - 4$

Exercise 7.2

Solve the following equations using the addition property

1-	$x + 1 = 0$		2-	$a - 2 = 0$
3-	$5 - b = 0$		4-	$6 - 2y = 0$
5-	$z - 9 = -2$		6-	$x - 11 = -3$
7-	$y - 3 = -15$		8-	$a - 7 = -7$

9-	$y + 2 = 8$	10-	$z - 4 = 14$
11-	$2a + 5 = 11$	12-	$3x - 3 = 12$
13-	$7y - 8 = 3y$	14-	$9z - 21 = 2z$
15-	$15a = -9 + 12a$	16-	$32x = 12x + 10$
17-	$y = -4 + 2y$	18-	$5b = -4b + 45$
19-	$6z + 1 = 11 + z$	20-	$x + 1 = 13 - x$
21-	$3a - 3 = 2a + 1$	22-	$4y - 1 = y + 17$
23-	$13b - 22 = 11b - 8$	24-	$8x - 30 = 3x - 5$
25-	$19z + 17 = 16z - 1$	26-	$7y - 2 = -3y + 38$
27-	$5x - 5 + x = 3 + 2x$	28-	$11a - 4 - 4a = 36 - a$

29- $\quad 23y - 11 + 2y = 9 + 15y$

30- $\quad 9x + 13 - 4x = 10x - 32$

31- $\quad 6b + 3 - 13b = 5 + 3b - 2$

32- $\quad 14x + 6 - 4x = -7 - 5x - 2$

33- $\quad 9y + 11 + 2y = -9 - 3y - 6y$

34- $\quad 35z + 3 - 22z = 39 - 2z - 3z$

35- $\quad 4(3x - 1) + 5(1 - 2x) = 13$

36- $\quad 3 + 2(4x - 9) = 5(2x - 1)$

37- $\quad 4(y + 3) - 6(2y + 1) = 10(y - 3)$

38- $\quad 24 + 3(4z - 2) = 6(4z - 1)$

39- $\quad 3a + 3(a - 2) = 5(a - 3) + 20$

40- $\quad 13 - 2(2b + 1) = 3(b - 1) - 2(b - 2)$

Exercise 7.3

Solve the following equations:

1- $0.02\,z = 0.32$ 2- $0.07\,y = 0.049$

3- $3.7\,x = 11.1$ 4- $2.2\,a = 15.4$

5- $1.9\,y = 7.6$ 6- $0.8\,y = -8$

7- $-0.3\,x = 3.6$ 8- $0.21\,b = 1.05$

9- $-0.11\,z + 0.23\,z = 0.72$ 10- $0.35\,b + 0.14\,b = 0.98$

11- $0.23\,a - 0.13\,a = 1.5$ 12- $-0.46\,y - 0.34\,y = -4.8$

13- $\dfrac{x}{7} = 10$ 14- $\dfrac{y}{11} = 3$

15- $-\dfrac{z}{3} = 9$ 16- $-\dfrac{a}{5} = -1$

17- $\dfrac{3}{4} \cdot b = 6$ 18- $\dfrac{5}{3} \cdot x = -10$

19- $\dfrac{5y}{-7} = 15$ 20- $\dfrac{-3a}{8} = -9$

21- $\dfrac{4z}{9} = \dfrac{8}{3}$ 22- $\dfrac{-22x}{5} = \dfrac{14}{-35}$

23- $\dfrac{a}{3} - \dfrac{1}{5} = \dfrac{1}{3} - \dfrac{a}{5}$ 24- $\dfrac{y}{4} - \dfrac{y}{6} = \dfrac{y}{2} + \dfrac{1}{2}$

25- $\dfrac{2z}{5} + \dfrac{3}{4} = \dfrac{4z}{5} + \dfrac{11}{4}$ 26- $\dfrac{2x}{3} + \dfrac{x}{4} - \dfrac{3x}{2} = -\dfrac{7}{2}$

27- $\dfrac{2b}{3} + \dfrac{5}{6} = \dfrac{3b}{4} + \dfrac{1}{6}$ 28- $\dfrac{9x}{2} + 10 = \dfrac{1}{2} - 5x$

29- $\dfrac{3y+3}{4} - \dfrac{6y+5}{12} = \dfrac{5}{6}$ 30- $\dfrac{z+1}{4} - \dfrac{7z+2}{7} = \dfrac{1}{4}$

127

31- $\dfrac{x-2}{5} - \dfrac{3-x}{4} = \dfrac{1}{5}$ 32- $\dfrac{a-2}{4} + \dfrac{2a+8}{3} = \dfrac{5}{4}$

33- $\dfrac{b+1}{3} + \dfrac{b-2}{3} = \dfrac{b}{2}$ 34- $\dfrac{2y-2}{7} - \dfrac{y-1}{2} = -\dfrac{9}{14}$

35- $\dfrac{x+7}{5} = \dfrac{2-x}{4} - \dfrac{x+2}{10}$ 36- $\dfrac{y-1}{3} = \dfrac{1-y}{6} + \dfrac{y+5}{9}$

37- $0.3\,(y + 10) - 0.1\,y = -0.2$

38- $0.4\,x - 0.5\,(6 - x) = 6$

39- $0.9\,z + 0.8\,(5 + z) = 10.8$

40- $0.75\,(a + 2) + 0.4\,(3\,a + 5) = -16$

Exercise 7.4

Solve the following word problems

1-   The sum of a number and 38 is 63. Find the number.

2-   If a number is multiplied by 3, then 12 is subtracted from the product, the answer is 30. Find the number.

3-   If one seventh of a number is subtracted from one third the same number, the answer is 4. Find the number.

4-   Two numbers, the first number is 4 times the second number, and the sum of the two numbers is 45. Find the two numbers.

5-   The sum of two numbers is 5. if we multiply the first number by 5 and add 2 to the product, the answer is 3 more than the second number multiplied by 3. Find the two numbers.

6-   A merchant bought a computer for $900. How much should the merchant sell it for to make 25% profit?

7-   George received a 5% raise this year, his salary now is $5040 per month what was George's salary before the raise?

8-   If you buy a refrigerator on sale, and the sale price was $1870. The sale price was 15% off of the original price. What was the original price of the refrigerator?

9-   Monty invested part of an amount of $60,000 at an interest rate of 6% per year, and the remainder of the $60,000 at an interest rate of 4.5% per year. If the total amount of interest he received from both investments after one year is $3,300. How much did he invest at each rate?

10-   Nabil invested two amounts of money at two different interest rates. The first amount was $32,000, and the second amount was $20,000. After one year, the amount of interest he received from the first amount was $700 more than the interest he received from the second investment. The interest rate of the second amount was 0.50% per year less than the interest rate of the first amount. Find the interest rate for each amount.

11-   Two planes are traveling in the same direction from the same line. The speed of the first plane is 380 miles per hours, and the speed of the second plane is 320 miles per hours. In how many hours will the two planes be 150 miles apart.

12-   Two ships traveling in the same direction. The first ship started the trip 2 hours before the second ship, and is navigating at a speed of 20 miles per hour. The second ship is navigating at a speed of 30 miles per hour. How many hours would the second ship take to be in line with the first ship?

13-   Two trains traveling in opposite directions toward each other, and they are 280 miles apart. The speed of the first train is 8 miles per hour more than the speed of the second train. If the two trains meet after 2 hours, find the speed of each train.

14-   George and John are riding their bicycles. They started from the same point heading the same direction to the movie theater. George's speed is 8 miles per hour, and reached the movie theater after 0.5 hour. What should John's speed be to arrive at the movie theater in 0.4 hour?

15-   Lydia had an appointment in Los Angeles, which is 70 miles from where she lives. Her one way trip took 1.5 hours. If Lydia drives at a speed of 60 miles per hour on the highway, and 20 miles per hour in the city, find the distance she drove in the city.

16-   Matilda's age is 4 times Jennifer's age. The sum of their ages is 115. Find their ages.

17-   If Vickie's age is multiplied by 3, then 9 is added to the product, the answer is 255. Find Vickie's age.

18-   The sum of Monty and Gigi's ages is 90. If five times Gigi's age equals four times Monty's age, find their ages.

19-   One third of Anthony's age is equal to one fifth of his age plus 4. Find Anthony's age.

20-   The sum of 3 students' ages is 62. The age of the second student is 4 less than the age of the first student. The age of the third student is 6 less than the age of the first student. Find the 3 students' ages.

21-   Mimi has $5.75 in quarters and dimes. If the total number of the coins is 35. How many coins of each does he have?

22-   Mike has $4.90 in dimes and nickels. If the number of dimes is 16 more than the number of nickels, how many coins of each does he have?

23-   Jack has $125 in ten and five dollar bills. The total number of bills is 16. How many bills of each does Jack have?

24- Reda has $300 in ten dollars, five dollars, and one dollar bills. The total number of bills is 98. The number of one dollar bills is 7 times the number of five dollar bills. How many bills of each does Reda have?

25- Laila has $10.20 in quarters, dimes, and nickels. She has twice as many quarters as dimes, and five times as many nickels as dimes. How many coins of each does Laila have?

26- The width of a rectangle is 11 feet less than its length. The perimeter of the rectangle is 30 feet. Find the length and width of the rectangle.

27- The first side of a triangle is 20 feet long. The second side is 3 times the length of the third side. If the perimeter of the triangle is 44 feet, find the length of the other two sides of the triangle.

28- The length of a rectangle is twice the width. If the width of the rectangle is 7 feet less than the length, find the length and width of the rectangle.

29- The perimeter of a square is ten times the length of the side minus 36. Find the length of the side of the square.

30- The circumference of any circle = $2 \pi r$ (where $\pi = 3.14$, and r is the radius of the circle). If the circumference of a circle is ten times the radius minus 11.16 feet, find the radius of the circle.

31- Two consecutive odd numbers. If the smaller number is multiplied by 4 and added to 2 times the larger number the sum is 58. Find the two numbers.

32-   Two consecutive even numbers. If the smaller number is
        multiplied by 3 and added to the larger number the sum is 90. Find
        the two numbers.

33-   The sum of three consecutive even numbers is 102. Find the three
        numbers.

34-   Three consecutive odd numbers. The sum of the first and third
        number equals the second number plus 27. Find the three
        numbers.

35-   Three consecutive even numbers. Two times the sum of the first
        and second numbers equals three times the third number plus 60.
        Find the three numbers.

36-   How many gallons of water must be added to 25 gallons of 72%
        iodine solution to make an iodine solution of 40% concentration?

37-   If you mix 10 gallons of 90% acid solution with 12 gallons of 68%
        acid solution, what will the acid concentration be in the resulting
        mixture?

38-   If you mix 14 pounds of 90% gold alloy with 6 pounds of 10%
        gold alloy, what will the gold percentage be in the resulting
        mixture?

39-   Mira mixed 50 gallons of 80% acid solution with 20 gallons of
        another acid solution. The resulting mixture had 60% acid
        concentration. What was the acid concentration in the second
        solution?

40-   Lydia mixed a 70% silver alloy with a 50% silver alloy, and the resulting mixture had 64% silver concentration. If the weight of the 70% silver alloy was 10 pounds more than the weight of the 50% silver alloy, what is the weight of the resulting mixture?

CHAPTER 8

Solving Quadratic Equations

Definition

A quadratic equation is an equation with one or more terms raised to the power 2. The standard form of a quadratic equation is $ax^2 + bx + c = 0$, where a, b, and c are constants (numbers). Solving quadratic equations yield two answers for the unknown. In a few cases, both answers are the same.

Factoring

Some quadratic equations can be factored to form linear equations. The resulting linear equations are solved using the methods explained in chapter 7. To make sure that the answers are correct, substitute each value obtained into the original equation to see if it yields the correct answer.

Example 1:   Solve the equation $y^2 - 9y = 0$

Factor y in the left side of the equation
$y(y - 9) = 0$

Now, equate each one of the two factors on the left side of the equation to 0, and find the value of y.
$y = 0$

Substitute 0 for y in the left side of the original equation
$0^2 - 9 \times 0 = 0 - 0 = 0$
$\therefore$ 0 is a correct answer

Equate the second factor on the left side of the equation to 0, and find the value of y.
$y - 9 = 0$

Add 9 to each side of the equation
$$y - 9 + 9 = 0 + 9$$
$$y = 9$$

Substitute 9 for y in the left side of the original equation
$$9^2 - 9 \times 9 = 81 - 81 = 0$$
$\therefore$ 9 is a correct answer

$\therefore$ y = 0, or y = 9

Example 2:   Solve the equation $9x^2 - 9x + 2 = 0$

Factor the left side of the equation
$$(3x - 2)(3x - 1) = 0$$

Equate each one of the two factors on the left side of the equation to 0, and find the value of x
$$3x - 2 = 0$$

Add 2 to each side of the equation
$$3x - 2 + 2 = 0 + 2$$
$$3x = 2$$

Divide each side of the equation by 3
$$x = \frac{2}{3}$$

Substitute $\frac{2}{3}$ for x in the original equation

$$9\left(\frac{2}{3}\right)^2 - 9\left(\frac{2}{3}\right) + 2$$
$$= 9\left(\frac{4}{9}\right) - 9\left(\frac{2}{3}\right) + 2$$
$$= 4 - 6 + 2 = 0$$
$\therefore \frac{2}{3}$ is a correct answer

Equate the second factor on the left side of the equation to 0, and find the value of x.

$3x - 1 = 0$

Add 1 to each side of the equation

$3x - 1 + 1 = 0 + 1$

$3x = 1$

Divide each side of the equation by 3

$$x = \frac{1}{3}$$

Substitute $\frac{1}{3}$ for x in the original equation

$$9\left(\frac{1}{3}\right)^2 - 9\left(\frac{1}{3}\right) + 2$$

$$= 9\left(\frac{1}{9}\right) - 9\left(\frac{1}{3}\right) + 2$$

$$= 1 - 3 + 2 = 0$$

$\therefore \frac{1}{3}$ is a correct answer

$$\therefore x = \frac{2}{3}, \text{ or } x = \frac{1}{3}$$

Example 3:   Solve the equation $y^2 - 16 = 0$

Factor the left side of the equation

$(y - 4)(y + 4) = 0$

Equate each one of the two factors on the left side of the equation to 0, and find the value of y.

$y - 4 = 0$

Add 4 to each side of the equation

$y - 4 + 4 = 0 + 4$

$y = 4$

Substitute 4 for y in the original equation
$(4)^2 - 16 = 16 - 16 = 0$
$\therefore$ 4 is a correct answer

Equate the second factor on the left side of the equation to 0, and find the value of y.
$y + 4 = 0$

Add -4 to each side of the equation
$y + 4 - 4 = 0 - 4$
$y = -4$

Substitute -4 for y in the original equation
$(-4)^2 - 16 = 16 - 16 = 0$
$\therefore$ -4 is a correct answer

$\therefore$ $y = 4$, or $y = -4$

Example 4:   Solve the equation $x^2 - 12x + 9 = 0$

Factor the left side of the equation

$(x - 3)(x - 3) = 0$
$x - 3 = 0$

Add 3 to each side of the equation
$x - 3 + 3 = 0 + 3$
$x = 3$

Similarly the other factor will yield $x = 3$ also

This example illustrates the fact that some quadratic equations have two equal solutions. When a quadratic equation has two equal solutions, the solutions are called **double root**.

Example 5:   Solve the equation $x^2 + 36 = 0$

Add -36 to each side of the equation
$$x^2 + 36 - 36 = 0 - 36$$
$$x^2 = -36$$

Take the square root of each side of the equation
$$x = \sqrt{-36}$$
Since the radicand is a negative number, the answer is an imaginary number equal to $\pm\, 6\, \boldsymbol{i}$

The following is an example of a quadratic equation that cannot be solved by factoring.

Example: Solve the equation $x^2 - 5x = 2$

Factor the left side of the equation
$$x\,(x - 5) = 2$$
$$x = 2$$

Substitute 2 for x in the original equation
$$2^2 - 5(2) = 4 - 10 = -6$$
Since the value of x did not yield 2 on the right side of the equation, then 2 is not a correct answer.

$$x - 5 = 2$$
Add 5 to each side of the equation
$$x - 5 + 5 = 2 + 5$$
$$x = 7$$

Substitute 7 for x in the original equation
$$7^2 - 5(7) = 49 - 35 = 14$$
Since the value of x did not yield 2 on the right side of the equation, then 7 is not a correct answer.

The main reason we get a wrong answer when we equate each term (x, and $x + 5$) To 2 is the fact that the two numbers multiplied times each other equals 2, not each one individually. However, when the right side of the equation is 0, we can equate each term individually to 0.

The question now is, how do we solve this type of quadratic equations, you will find the answer to this question later on in this chapter, after you are introduced to the different methods used to solve quadratic equations.

Completing the Square

A closer look at perfect squares in the form of $(x + c)^2$ which equals $x^2 + 2cx + c^2$, and the form $(x - c)^2$ which equals $x^2 - 2cx + c^2$ tells us that if an equation includes only the terms $x^2 + 2cx$, or $x^2 - 2cx$ we can make this equation a perfect square if we add the term c^2.
Before we start to calculate the term c^2 we have to make sure that the coefficient of x^2 equals 1. If it is not, divide all the terms of the equation by that coefficient.

To calculate the term c^2, divide the coefficient of x (which is $2c$) by 2 (this results in c), then square it (it becomes c^2). To make sure that we do not change the value of the original equation, this term (c^2) must be added to both sides of the equation.

This method can be used to solve quadratic equations in the form of:
$x^2 + 2cx = b$, which can also be written in the form $x^2 + 2cx - b = 0$,
or $x^2 - 2cx = b$, which can also be written in the form $x^2 - 2cx - b = 0$,
where c is a constant other than zero, and b is zero or any other positive or negative number.

Example 1:   Solve the following equation by completing the square
$$x^2 - 2x = 0$$

Divide the coefficient of x by 2
$$\frac{-2}{2} = -1$$

Square the quotient
$$(-1)^2 = 1$$

Add this number (1) to each side of the equation
$$x^2 - 2x + 1 = 0 + 1$$
$$x^2 - 2x + 1 = 1$$
$$(x - 1)^2 = 1$$

Take the square root of each side of the equation

$x - 1 = 1$ or $x - 1 = -1$

$\therefore x = 2$ or $x = 0$

Example 2:   Solve the following equation by completing the square
$$2x^2 - 8x = 0$$

Divide all the terms of the equation by 2
$$x^2 - 4x = 0$$

Divide the coefficient of x by 2
$$\frac{-4}{2} = -2$$

Square the quotient
$$(-2)^2 = 4$$

Add this number (4) to each side of the equation
$$x^2 - 4x + 4 = 0 + 4$$
$$x^2 - 4x + 4 = 4$$
$$(x - 2)^2 = 4$$

Take the square root of each side of the equation

$x - 2 = 2$ or $x - 2 = -2$

$\therefore x = 4$ or $x = 0$

140

Example 3:   Solve the following equation by completing the square
$$x^2 + 10\ x = 11$$

Divide the coefficient of x by 2
$$\frac{10}{2} = 5$$

Square the quotient
$$(5)^2 = 25$$

Add this number (25) to each side of the equation
$$x^2 + 10\ x + 25 = 11 + 25$$
$$x^2 + 10\ x + 25 = 36$$
$$(x + 5)^2 = 36$$

Take the square root of each side of the equation
$$x + 5 = 6 \qquad \text{or} \quad x + 5 = -6$$
$$\therefore x = 1 \qquad \text{or} \quad x = -11$$

Example 4:   Solve the following equation by completing the square
$$3x^2 - 36\ x = -60$$

Divide all the terms of the equation by 3
$$x^2 - 12\ x = -20$$

Divide the coefficient of x by 2
$$\frac{-12}{2} = -6$$

Square the quotient
$$(-6)^2 = 36$$

Add this number (36) to each side of the equation
$$x^2 - 12\ x + 36 = -20 + 36$$

141

$$x^2 - 12x + 36 = 16$$
$$(x - 6)^2 = 16$$

Take the square root of each side of the equation

$$x - 6 = 4 \quad \text{or} \quad x - 6 = -4$$
$$\therefore x = 10 \quad \text{or} \quad x = 2$$

Quadratic Formula

The quadratic formula can be used to solve any quadratic equation in the form of $ax^2 + bx + c = 0$, where (a) does not equal zero.

Quadratic formula .... $x = \dfrac{-b \pm \sqrt{b^2 - 4ac}}{2a}$

In other words, the two values of x are:

$$x = \dfrac{-b + \sqrt{b^2 - 4ac}}{2a}$$

$$\text{or } x = \dfrac{-b - \sqrt{b^2 - 4ac}}{2a}$$

where (a) is the coefficient of x^2, (b) is the coefficient of x, and (c) is the constant number. (a), (b), and (c) can be either positive or negative numbers.

Example 1:   Solve the following equation using the quadratic formula
$$x^2 - 5x + 6 = 0$$

$$a = 1 \qquad b = -5 \qquad c = 6$$

$$x = \dfrac{-b \pm \sqrt{b^2 - 4ac}}{2a}$$

$$x = \frac{-(-5) \pm \sqrt{(-5)^2 - 4(1)(6)}}{2(1)}$$

$$x = \frac{5 \pm \sqrt{25 - 24}}{2}$$

$$x = \frac{5 \pm \sqrt{1}}{2}$$

$$x = \frac{5 \pm 1}{2}$$

$$x = \frac{5 + 1}{2} = \frac{6}{2} = 3$$

$$\text{or } x = \frac{5 - 1}{2} = \frac{4}{2} = 2$$

$\therefore$ the two values of x are $(3, 2)$

Example 2: Solve the following equation using the quadratic formula
$x^2 - x - 12 = 0$

$$a = 1 \qquad b = -1 \qquad c = -12$$

$$x = \frac{-b \pm \sqrt{b^2 - 4ac}}{2a}$$

$$x = \frac{-(-1) \pm \sqrt{(-1)^2 - 4(1)(-12)}}{2(1)}$$

$$x = \frac{-1 \pm \sqrt{1 + 48}}{2}$$

$$x = \frac{-1 \pm \sqrt{49}}{2}$$

$$x = \frac{-1 \pm 7}{2}$$

$$x = \frac{-1 + 7}{2} = \frac{6}{2} = 3$$

$$\text{Or } x = \frac{-1 - 7}{2} = \frac{-8}{2} = -4$$

$\therefore$ the two values of x are $(3, -4)$

Example 3: Solve the following equation using the quadratic formula
$4x^2 + 32x + 60 = 0$

$a = 4$ $\qquad$ $b = 32$ $\qquad$ $c = 60$

$$x = \frac{-b \pm \sqrt{b^2 - 4ac}}{2a}$$

$$x = \frac{-32 \pm \sqrt{32^2 - 4(4)(60)}}{2(4)}$$

$$x = \frac{-32 \pm \sqrt{1024 - 960}}{8}$$

$$x = \frac{-32 \pm \sqrt{64}}{8}$$

144

$$x = \frac{-32 \pm 8}{8}$$

$$x = \frac{-32 + 8}{8} = \frac{-24}{8} = -3$$

$$x = \frac{-32 - 8}{8} = \frac{-40}{8} = -5$$

$\therefore$ the two values of x are $(-3, -5)$

Example 4:   Solve the following equation using the quadratic formula
$9x^2 - 9x - 18 = 0$

a = 9   b = -9   c = -18

$$x = \frac{-b \pm \sqrt{b^2 - 4ac}}{2a}$$

$$x = \frac{-(-9) \pm \sqrt{(-9)^2 - 4(9)(-18)}}{2(9)}$$

$$x = \frac{9 \pm \sqrt{81 + 648}}{18}$$

$$x = \frac{9 \pm \sqrt{729}}{18}$$

$$x = \frac{9 \pm 27}{18}$$

$$x = \frac{9 + 27}{18} = \frac{36}{18} = 2$$

145

or $x = \dfrac{9 - 27}{18} = \dfrac{-18}{18} = -1$

$\therefore$ the two values of x are $(2, -1)$

Example 5: Solve the following equation using the quadratic formula
$4x^2 - 32x + 64 = 0$

To simplify the equation divide all terms by 4
$x^2 - 8x + 16 = 0$

$a = 1 \qquad\qquad b = -8 \qquad\qquad c = 16$

$$x = \dfrac{-b \pm \sqrt{b^2 - 4ac}}{2a}$$

$$x = \dfrac{-(-8) \pm \sqrt{(-8)^2 - 4(1)(16)}}{2(1)}$$

$$x = \dfrac{8 \pm \sqrt{64 - 64}}{2}$$

$$x = \dfrac{8 \pm 0}{2}$$

$$x = \dfrac{8}{2} = 4$$

or $x = \dfrac{8}{2} = 4$

$\therefore$ the two values of x are $(4, 4)$

Special Cases

In the cases when one of the terms is missing from a quadratic equation, it is easier and faster to solve these types of equations using other methods such as factoring or taking the square root of each side of the equation.

Example 1:   Solve the equation $5x - 20 = 0$

As explained before, the standard form of a quadratic equation is $ax^2 + bx + c = 0$
It is obvious that the equation we are trying to solve is not a quadratic equation because it is missing the term ax^2, and we cannot use the quadratic formula to solve it.

To solve this equation add 20 to each side of the equation
$5x - 20 + 20 = 0 + 20$
$5x = 20$
$x = 4$

Example 2:   Solve the equation $5x^2 - 20 = 0$

This is a quadratic equation missing the term bx.
We can solve this equation using the quadratic formula; however, it can also be solved as follows:

Add 20 to each side of the equation
$5x^2 - 20 + 20 = 0 + 20$
$5x^2 = 20$
$x^2 = 4$

Take the square root of each side of the equation
$x = 2$ or $x = -2$

Example 3:   Solve the equation $4x^2 + 8\ x = 0$

This is a quadratic equation missing the term c.
We can solve this equation using the quadratic formula;
however, it can also be solved by factoring as follows:

Factor $4x$

$$4x(x+2) = 0$$

Equate each one of the terms on the left side of the equation
to 0.

$$4x = 0$$

$$x = 0$$

or $(x+2) = 0$

$$x = -2$$

Exercise 8.1

Solve the following equations by factoring:

1-	$x^2 - 2x = 0$	2-	$x^2 - 5x = 0$
3-	$3x^2 - 12x = 0$	4-	$3x^2 + 21x = 0$
5-	$4x^2 + 4x = 0$	6-	$4x^2 - 12x = 0$
7-	$2x^2 - 4x = 0$	8-	$2x^2 + 5x = 0$
9-	$6x^2 + 18x = 0$	10-	$4x^2 + 6x = 0$
11-	$9x^2 + 15x = 0$	12-	$8x^2 - 36x = 0$
13-	$12x^2 - 33x = 0$	14-	$5x^2 - 30x = 0$

15- $10x^2 - 25x = 0$ 16- $4x^2 + 6x = 0$

17- $6x^2 + 16x = 0$ 18- $8x^2 - 18x = 0$

19- $15x^2 + 35x = 0$ 20- $20x^2 - 55x = 0$

Exercise 8.2

Solve the following equations:

(Hint: rearrange the equation to have the term (x^2) on the left side of the equation and the other term on the right side of the equation).

1- $x^2 - 4 = 0$ 2- $x^2 - 16 = 0$

3- $x^2 - 25 = 0$ 4- $x^2 - 49 = 0$

5- $4x^2 - 9 = 0$ 6- $2x^2 - 14 = 0$

7- $2x^2 - 3 = 0$ 8- $4x^2 - 3 = 0$

9- $3x^2 - 18 = 0$ 10- $3x^2 - 147 = 0$

11- $5x^2 - 10 = 0$ 12- $5x^2 - 25 = 0$

13- $2x^2 - 18 = 0$ 14- $2x^2 - 25 = 0$

15- $3x^2 - 9 = 0$ 16- $3x^2 - 16 = 0$

17- $4x^2 - 20 = 0$ 18- $4x^2 - 32 = 0$

19- $5x^2 - 16 = 0$ 20- $5x^2 - 100 = 0$

21- $x^2 + 4 = 0$ 22- $x^2 + 16 = 0$

23- $2x^2 + 18 = 0$ 24- $2x^2 + 50 = 0$

Exercise 8.3

Factor the following terms by completing the square:

1- $x^2 - 6x = 0$ 2- $x^2 - 8x = 0$

3-	$x^2 - 10x = 0$	4-	$x^2 - 12x = 0$
5-	$x^2 - 14x = 0$	6-	$x^2 + 6x = 0$
7-	$x^2 + 8x = 0$	8-	$x^2 + 10x = 0$
9-	$x^2 + 12x = 0$	10-	$x^2 + 14x = 0$
11-	$x^2 - 3x = 0$	12-	$x^2 - 5x = 0$
13-	$x^2 - 7x = 0$	14-	$x^2 - 9x = 0$
15-	$x^2 - 11x = 0$	16-	$x^2 - \dfrac{3}{2}x = 0$
17-	$x^2 - \dfrac{2}{3}x = 0$	18-	$x^2 - \dfrac{1}{2}x = 0$
19-	$x^2 - \dfrac{2}{5}x = 0$	20-	$x^2 - \dfrac{4}{7}x = 0$
21-	$x^2 - \dfrac{4}{3}x = 0$	22-	$x^2 - \dfrac{4}{5}x = 0$
23-	$y^2 - 10y + 16 = 0$	24-	$x^2 + 4x - 12 = 0$
25-	$z^2 - 6z + 1 = 0$	26-	$4y^2 - 20y + 24 = 0$
27-	$4a^2 + 4a - 3 = 0$	28-	$z^2 - 2z - 8 = 0$
29-	$3b^2 + b - 3 = 0$	30-	$3y^2 - 2y - 1 = 0$
31-	$y^2 + 2y = 4$	32-	$a^2 + 4a = 9$
33-	$z^2 + 6z = 16$	34-	$b^2 + 8b = 33$
35-	$y^2 + 10y = 39$	36-	$x^2 = 2x + 4$
37-	$a^2 = 4a + 4$	38-	$z^2 = 6z + 16$
39-	$b^2 = 8b + 33$	40-	$x^2 = 10x + 39$

Exercise 8.4

Solve the following equations using the quadratic formula:

1-	$x^2 - 16 = 0$	2-	$x^2 - 25 = 0$
3-	$x^2 - 36 = 0$	4-	$x^2 - 49 = 0$
5-	$x^2 - 64 = 0$	6-	$x^2 - x = 0$
7-	$x^2 - 2x = 0$	8-	$x^2 - 3x = 0$
9-	$x^2 - 4x = 0$	10-	$x^2 - 5x = 0$
11-	$2y^2 - 4y = 0$	12-	$2y^2 - 6y = 0$
13-	$2y^2 - 8y = 0$	14-	$2y^2 - 10y = 0$

15-	$2y^2 - 12y = 0$	16-	$3y^2 - 2y = 0$
17-	$3y^2 - 4y = 0$	18-	$4y^2 - y = 0$
19-	$4y^2 - 2y = 0$	20-	$4y^2 - 5y = 0$
21-	$x^2 - 3x + 2 = 0$	22-	$x^2 - 5x + 4 = 0$
23-	$x^2 - 7x + 12 = 0$	24-	$x^2 - 6x + 6 = 0$
25-	$x^2 - 6x + 5 = 0$	26-	$x^2 - 7x + 10 = 0$
27-	$x^2 - 9x + 20 = 0$	28-	$x^2 - 8x + 15 = 0$
29-	$x^2 - 2x + 1 = 0$	30-	$x^2 - 6x + 9 = 0$
31-	$y^2 - 10y + 24 = 0$	32-	$y^2 - 7y + 6 = 0$
33-	$y^2 - 8y + 12 = 0$	34-	$y^2 - 9y + 18 = 0$
35-	$y^2 - 11y + 30 = 0$	36-	$y^2 - 10y + 25 = 0$
37-	$y^2 - 12y + 36 = 0$	38-	$y^2 - 4y + 3 = 0$
39-	$y^2 + y - 30 = 0$	40-	$y^2 - 5y + 6 = 0$
41-	$2x^2 + 5x + 2 = 0$	42-	$2x^2 + 8x + 6 = 0$
43-	$3x^2 + 9x - 12 = 0$	44-	$3x^2 + 9x - 30 = 0$
45-	$3x^2 + 14x - 24 = 0$	46-	$4x^2 + 18x + 20 = 0$
47-	$9x^2 - 9x - 18 = 0$	48-	$10x^2 + 5x - 15 = 0$
49-	$6x^2 + 27x + 30 = 0$	50-	$8x^2 + 12x - 36 = 0$

CHAPTER 9

INEQUALITIES

Definition

A mathematical expression is a set of real numbers, unknowns, or a combination thereof. An inequality is defined as a statement indicating that a mathematical expression is one or more of the following:

1-   Greater than another mathematical expression.
      Example: $x > 4$

2-   Less than another mathematical expression.
      Example: $x < 7$

3-   Greater than or equal to another mathematical expression.
      Example: $x \geq -2$

4-   Less than or equal to another mathematical expression.
      Example: $x \leq 5$

5-   A combination of the above.

      Example 1: $9 > x > 6$

      Example 2: $8 \geq x \geq 5$

      Example 3: $7 > x \geq 2$

      Example 4: $4 \geq x > 1$

What does the answer mean?

When we solve an **equation** the answer we get is usually one or more real numbers, which include rational numbers, or irrational numbers, and can be positive, negative or zero. For example $x = 3$ means that x has only **one value**, which is exactly 3.

However, when we solve an **inequality** the answer we get is not one specific exact value, rather, the answer represents a set of numbers. For example, if the answer is $x > 4$ this means that the value of x is **all the real numbers** greater than 4.

In some cases, inequality statements are not true for any real number. For example the statement $2x + 3 > 2x + 8$ is not true for any real number. Also the statement $x + 5 \leq x + 1$ is not true for any real number.

Solving Linear Inequalities in one variable

To solve linear inequalities in one variable we can use the addition property, subtraction property, multiplication property (positive numbers), or the division property (positive numbers).

Example 1: Solve the inequality $x + 3 > 10$

Add -3 to each side of the inequality
$x + 3 - 3 > 10 - 3$
$x > 7$

Example 2: Solve the inequality $2x - 5 < 15$

Add $+5$ to each side of the inequality
$2x - 5 + 5 < 15 + 5$
$2x < 20$
Divide each side of the inequality by 2
$x < 10$

Example 3:   Solve the inequality $9 - 2x \geq 12 - 3x$

Add $3x$ to each side of the inequality
$9 - 2x + 3x \geq 12 - 3x + 3x$
$9 + x \geq 12$

Add -9 to each side of the inequality
$9 + x - 9 \geq 12 - 9$
$x \geq 3$

Example 4:   Solve the inequality $6x + 5 \leq 4x + 13$

Add $-4x$ to each side of the inequality
$6x + 5 - 4x \leq 4x + 13 - 4x$
$2x + 5 \leq 13$

Add -5 to each side of the inequality
$2x + 5 - 5 \leq 13 - 5$
$2x \leq 8$

Divide each side of the inequality by 2
$x \leq 4$

Multiplying and dividing an inequality by a negative number

As explained above, when we multiply each side of an inequality by a positive number the value of the inequality remains the same. The same is true when we divide each side of an inequality by a positive number.

However, when we multiply each side of an inequality by a negative number or divide each side of an inequality by a negative number, the value of the inequality is reversed, for example greater than (>) becomes less than (<), and greater than or equal to ($\geq$) becomes less than or equal to ($\leq$) .

Example 1: $2 < 4$

Multiply each side of the inequality by -5
$2\,(-5) < 4\,(-5)$
We get $-10 < -20$, which is not a true statement.

To correct the answer, we have to reverse the symbol between the numbers.
The correct answer is $-10 > -20$

Example 2: $21 > 7$

Divide each side of the inequality by -7
$$\frac{21}{-7} > \frac{7}{-7}$$
We get $-3 > -1$, which is not a true statement.
To correct the answer, we have to reverse the symbol between the numbers.
The correct answer is $-3 < -1$

Solving Inequalities that include absolute values

The absolute value of an unknown such as x, written $\lvert\, x \,\rvert$,
is equal to x, if x is ≥ 0
or is equal to $-x$, if x is < 0

Inequalities that include absolute values are solved using the steps shown in the examples below.

Example 1:   Find the values of x in the following inequality:

$$\lvert\, 2x - 4 \,\rvert \geq 8$$

Since the value of $2x - 4$ may be ≥ 0, or it may be < 0, we have to consider both of these two possibilities to find the values of x.

FIRST: If $2x - 4 \geq 0$, then $x \geq 2$

The absolute value of $2x - 4$, which is written as $|2x - 4|$, equals $2x - 4$

Now we can write the original equation as:

$2x - 4 \geq 8$

Add 4 to each side of the inequality

$2x - 4 + 4 \geq 8 + 4$

$2x \geq 12$

Which yields a value of $x \geq 6$

SECOND: If $2x - 4 < 0$, then $x < 2$

In this case the absolute value of $2x - 4$, which is written as $|2x - 4|$, equals $-(2x - 4) = -2x + 4$

Now we can write the original equation as:

$-2x + 4 \geq 8$

Multiply both sides of the equation by -1 and reverse the symbol from $\geq$ to $\leq$

$2x - 4 \leq -8$

Add 4 to each side of the inequality

$2x - 4 + 4 \leq -8 + 4$

$2x \leq -4$

$x \leq -2$

The two values of x are ≥ 6 and ≤ -2

Example 2:   Find the values of x in the following inequality:

$$|4x - 3| \leq x - 6$$

Since the value of $4x - 3$ may be ≥ 0, or it may be < 0, we have to consider both of these two possibilities to find the values of x.

FIRST: If $4x - 3 \geq 0$, then $x \geq \dfrac{3}{4}$

The absolute value of $4x - 3$, which is written as $|4x - 3|$, equals $4x - 3$

Now we can write the original equation as:

$$4x - 3 \leq x - 6$$

Add 3 to each side of the inequality

$$4x - 3 + 3 \leq x - 6 + 3$$

$$4x \leq x - 3$$

Add $- x$ to each side of the inequality

$$4x - x \leq x - 3 - x$$

$$3x \leq -3$$

Which yields a value of $x \leq -1$

SECOND: If $4x - 3 < 0$, then $x < \dfrac{3}{4}$

In this case the absolute value of $4x - 3$, which is written as $|4x - 3|$, equals $-(4x - 3) = -4x + 3$

Now we can write the original equation as:

$$-4x + 3 \leq x - 6$$

157

Add $-x$ to each side of the inequality

$-4x + 3 - x \leq x - 6 - x$

$-5x + 3 \leq -6$

Add -3 to each side of the inequality

$-5x + 3 - 3 \leq -6 - 3$

$-5x \leq -9$

Multiply each side of the inequality by -1 and reverse the symbol from $\leq$ to $\geq$

$5x \geq 9$

$x \geq \dfrac{9}{5}$

The two values of x are ≤ -1 and $\geq \dfrac{9}{5}$

Exercise 9.1

Solve the following inequalities:

1-	$x + 1 < 4$	2-	$x - 1 > 2$
3-	$x + 2 < 6$	4-	$x - 2 > 3$
5-	$x - 4 \leq 7$	6-	$x + 5 \geq 6$
7-	$x - 3 \geq -4$	8-	$x + 7 \leq -3$
9-	$2x - 7 < -13$	10-	$3x - 9 > -11$
11-	$2x - 5 > x + 1$	12-	$3x + 2 < x + 6$
13-	$5x - 1 > x + 2$	14-	$4x - 3 < x + 9$
15-	$11x + 1 \geq 2x + 10$	16-	$4x + 12 \leq x - 15$
17-	$7x - 2 > 1 + 10x$	18-	$2x + 3 < 4 + 3x$
19-	$3x + 7 > 3 + 7x$	20-	$6x + 2 < 6 + 9x$
21-	$8x - 7 \geq 2 + 11x$	22-	$6x + 1 \leq 5 + 7x$
23-	$7x - 6 + 2x \geq 10$	24-	$2x + 3 - 9x \leq 11$

25-	$4x - 3 - 2x > 5$	26-	$3x - 5 - x < 7$
27-	$9x + 1 - x > 17$	28-	$10x + 4 - 3x < 18$
29-	$3x + 2 - 2x \geq 1$	30-	$5x - 3 - x \leq 13$
31-	$7x - x + 1 \geq x + 11$	32-	$11x - 3x - 1 \leq 2x + 2$
33-	$12x - x - 1 \geq 3x - 9$	34-	$6x + x + 2 \leq x - 4$
35-	$8x + 3 + x > 9 + 3x$	36-	$4x + 9 + 3x < 19 + 2x$
37-	$5x + 7 - x > 9x - 10$	38-	$2x - 3 - x < 6x + 9$
39-	$3x + 2 - x \geq 5x - 11$	40-	$10x + 4 - x \leq -11 - 3x$

Exercise 9.2

Solve the following inequalities:

1-	$|x - 2| > 4$	2-	$|x - 1| < 2$
3-	$|x - 2| > 7$	4-	$|x + 5| < 3$
5-	$|x + 8| \geq 11$	6-	$|x - 6| \leq 5$
7-	$|x - 3| > 6$	8-	$|x - 7| < 1$
9-	$|x - 5| > -1$	10-	$|x + 6| < -2$
11-	$|2x - 4| \geq 6$	12-	$|2x + 6| \leq 9$
13-	$|3x - 2| \geq -8$	14-	$|3x - 1| \leq -13$
15-	$|4x + 3| \geq 17$	16-	$|5x - 2| \leq -7$
17-	$|6x - 1| > 2$	18-	$|4x - 5| < -8$
19-	$|7x + 1| > -9$	20-	$|3x + 2| < -11$
21-	$|5x - 3| \geq 1$	22-	$|8x + 3| \leq -2$
23-	$|10x + 4| > 5$	24-	$|9x - 4| < 4$

25- $\left| 2 - x \right| > 2x + 3$ 26- $\left| 2x - 6 \right| < x - 3$

27- $\left| 2x + 3 \right| > x + 9$ 28- $\left| 2x + 4 \right| < x - 5$

29- $\left| x - 10 \right| \geq 10 - x$ 30- $\left| 3x + 4 \right| \leq 6 - 7x$

31- $\left| 4x + 1 \right| > 2 - x$ 32- $\left| 5x + 3 \right| < 8 - x$

33- $\left| 3x - 3 \right| > 7 - 2x$ 34- $\left| 6x + 2 \right| < 5 - 3x$

35- $\left| 4x - 6 \right| \geq 3 - 5x$ 36- $\left| 5x + 1 \right| < 11 - 2x$

37- $\left| 7x - 1 \right| \leq 1 - 3x$ 38- $\left| 8x - 5 \right| > 3 - 2x$

39- $\left| 11x - 4 \right| < 1 - x$ 40- $\left| 2x - 7 \right| > -9 - 7x$

CHAPTER 10

GRAPHING LINEAR EQUATIONS

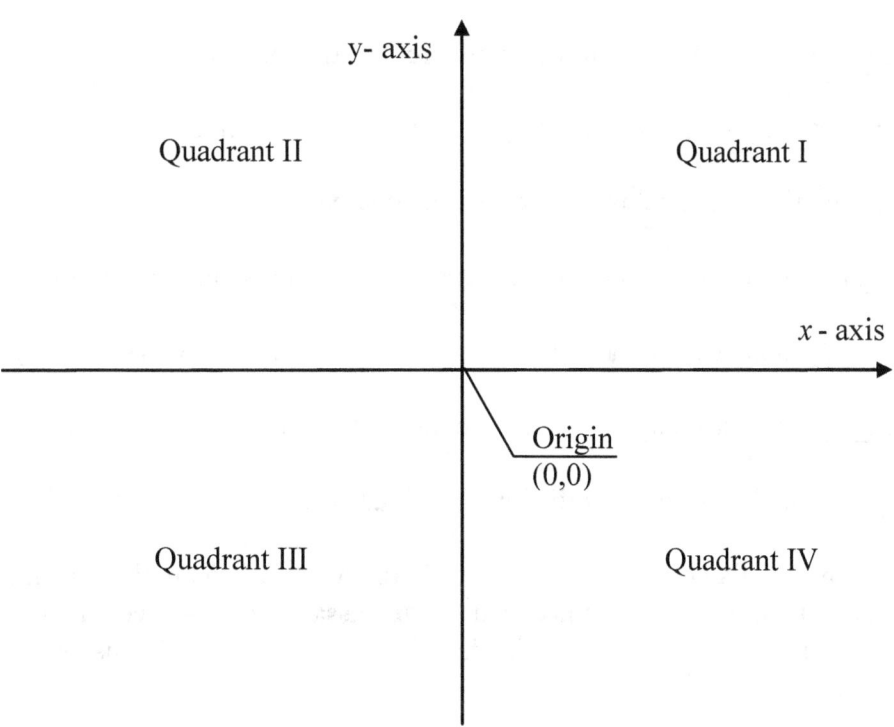

Figure 10.1

Coordinates

The **Cartesian System**, which is also called the **rectangular** system, provides a method to locate points using a horizontal line called the x-axis, and a vertical line called the y-axis. The point where the x and y-axis intersect is called the origin. The two axis x and y form four **quadrants**, see figure 10.1.

All points on the x-axis right of the origin are positive numbers.

All points on the x-axis left of the origin are negative numbers.

All points on the y-axis above the origin are positive numbers.

All points on the y-axis below the origin are negative numbers.

Quadrant 1: includes only positive x-coordinates, and positive y-coordinates.

Quadrant 2: includes only negative x-coordinates, and positive y-coordinates.

Quadrant 3: includes only negative x-coordinates, and negative y-coordinates.

Quadrant 4: includes only positive x-coordinates, and negative y-coordinates.

The coordinates of a point (x, y) are given in the form of two numbers such as (3, 2). The first number, which is called the **abscissa**, is the distance on the x-axis, and the second number, which is called the ordinate, is the distance on or along the y-axis, see figure 10.2.

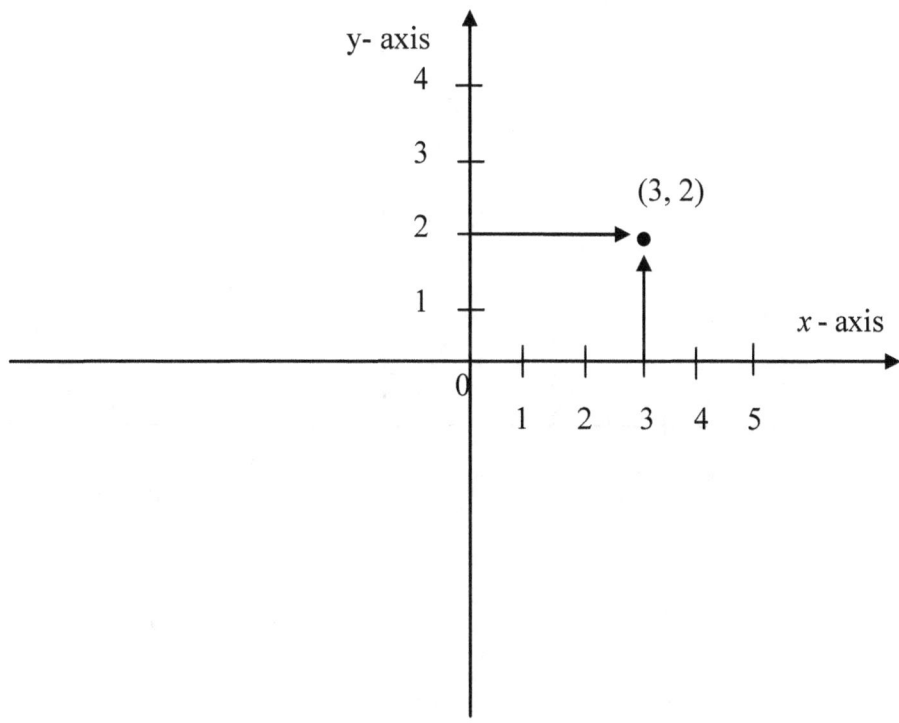

Figure 10.2

Figure 10.3 shows the location of the points $(-3, 3)$, $(-5, -4)$, $(4, -3)$.

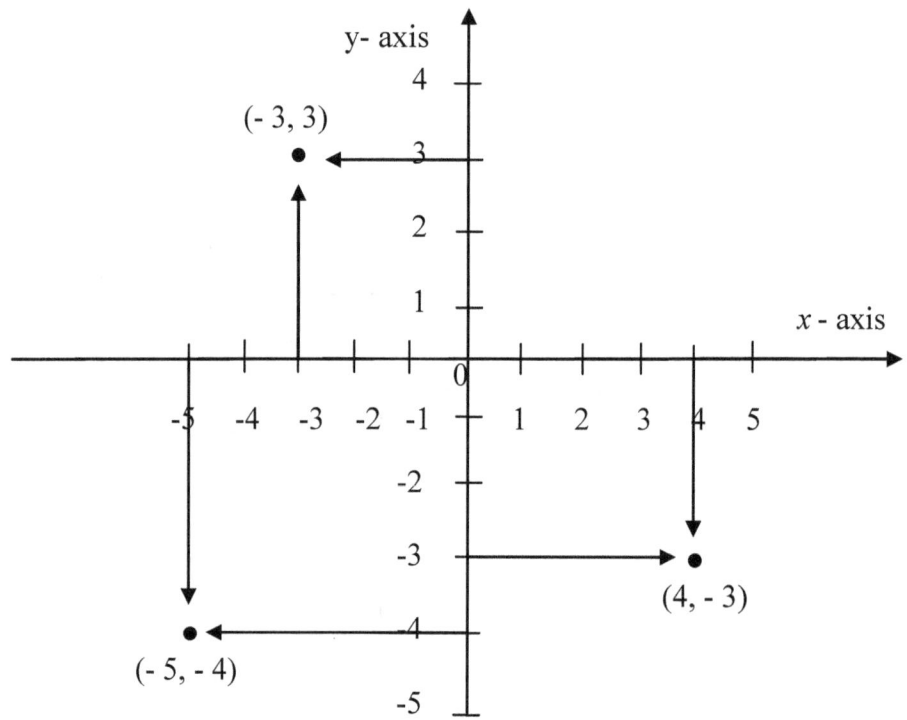

Figure 10.3

Graphing Linear Equations

Equations in two variables such as x and y are called linear equations. The graph of a linear equation is a **straight line**. The coordinates (x, y), which are also called the **ordered pair**, of all the points that satisfy a linear equation are called a **solution set**.

The general form of a linear equation is .... $ax + by = c$
a and b cannot be both zero.
The solution set of a linear equation includes an infinite number of ordered pairs.
To graph a linear equation in two variables all we need is only two ordered pairs
from the solution set.

We start by assuming a value, any value, of x, then substitute this value in the
linear equation, and calculate the value of y. The values of x and y form an
ordered pair from the solution set.

Repeat the process of assuming another value of x or y, then substitute this value
in the linear equation, and calculate the value of y or x. The values of x and y
form another ordered pair from the solution set.

Example 1: Find two ordered pairs from the solution set of the equation
$2x + 2y = 6$, and graph the equation.

Assume $x = 0$
Substitute 0 for the value of x in the equation

$2(0) + 2y = 6$
$2y = 6$
$y = 3$

The first ordered pair is (0, 3)

Assume $x = 2$
Substitute 2 for the value of x in the equation

$2(2) + 2y = 6$
$4 + 2y = 6$
Add -4 to each side of the equation
$4 + 2y - 4 = 6 - 4$
$2y = 2$
$y = 1$

The second ordered pair is (2, 1)

Plot the two ordered pairs, and connect them with a straight line. This straight line is the graph of the equation.
See figure 10.4

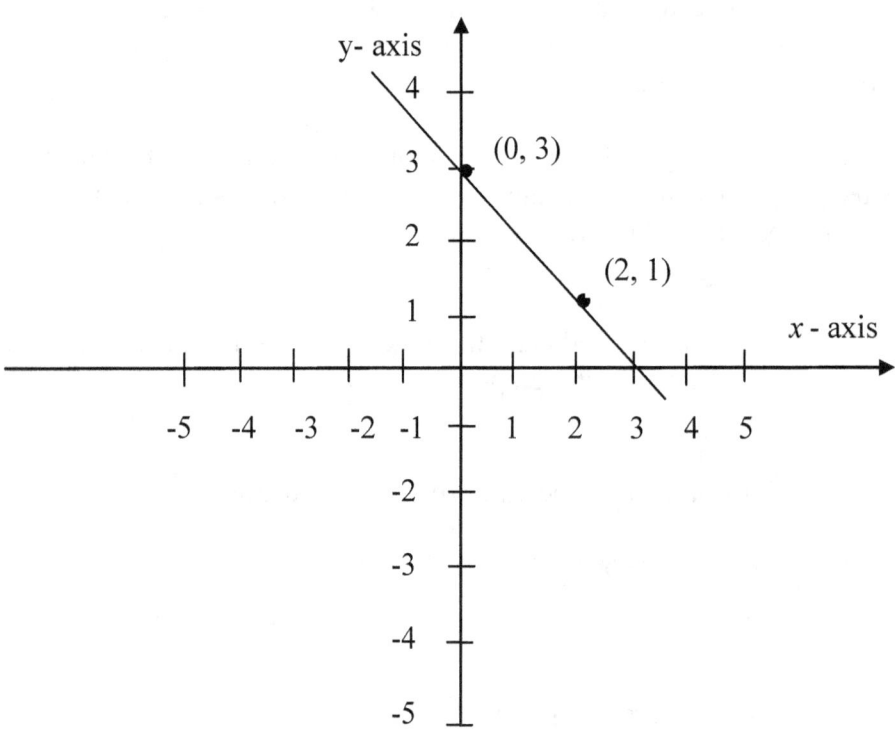

Figure 10.4

Example 2:   Find two ordered pairs from the solution set of the equation $x - 2y = -4$, and graph the equation.

Assume $x = 0$
Substitute 0 for the value of x in the equation

$(0) - 2\,y = -4$

Multiply each side of the equation by -1
$2\,y = 4$
$y = 2$

The first ordered pair is $(0, 2)$

Assume $y = 0$
Substitute 0 for the value of y in the equation

$x - 2(0) = -4$
$x = -4$

The second ordered pair is $(-4, 0)$

Plot the two ordered pairs, and connect them with a straight line. This straight line is the graph of the equation.
See figure 10.5

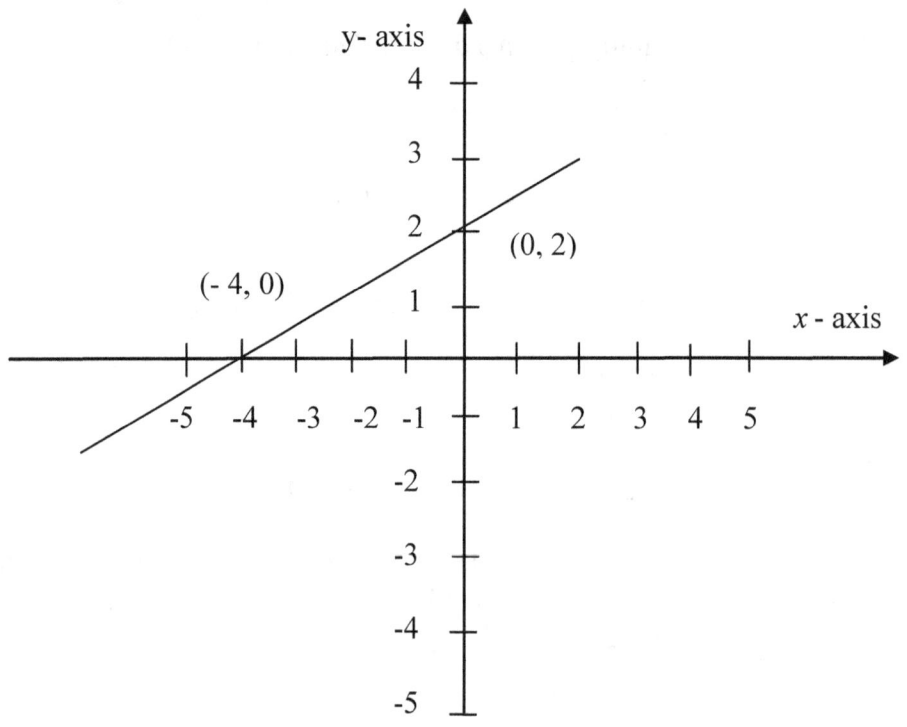

Figure 10.5

Special Linear Equations

When a linear equation include only one unknown such as $x = 4$, or $y = 3$, the equation is still a linear equation that represents a straight line, however, the straight line is a vertical line parallel to the y-axis if the equation is in the form of $x = c$, where c is 0, a positive number, or a negative number. $x = 0$, is the equation of the y-axis. When $x =$ a positive number, this is the equation of a vertical line at a distance (c) right of the y-axis. When $x =$ a negative number, this is the equation of a vertical line at a distance (c) left of the y-axis.

See figure 10.6

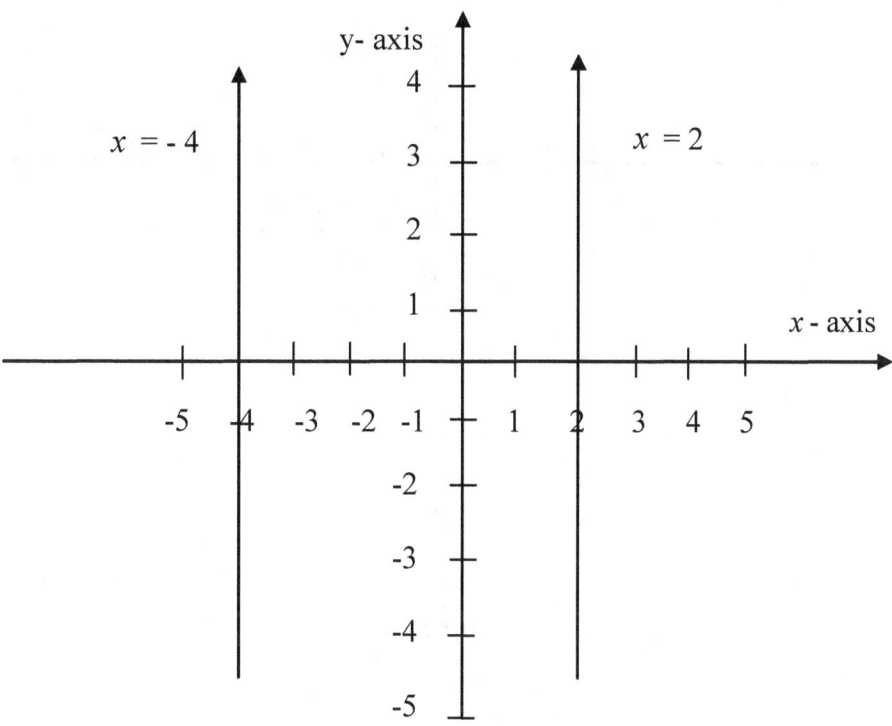

Figure 10.6

Similarly, if the equation is in the form of $y = c$, where c is 0, a positive number, or a negative number. $y = 0$, is the equation of the x - axis. When $y = $ a positive number, this is the equation of a horizontal line at a distance (c) above the x - axis. When $y = $ a negative number, this is the equation of a horizontal line at a distance (c) below the x - axis. See figure 10.7

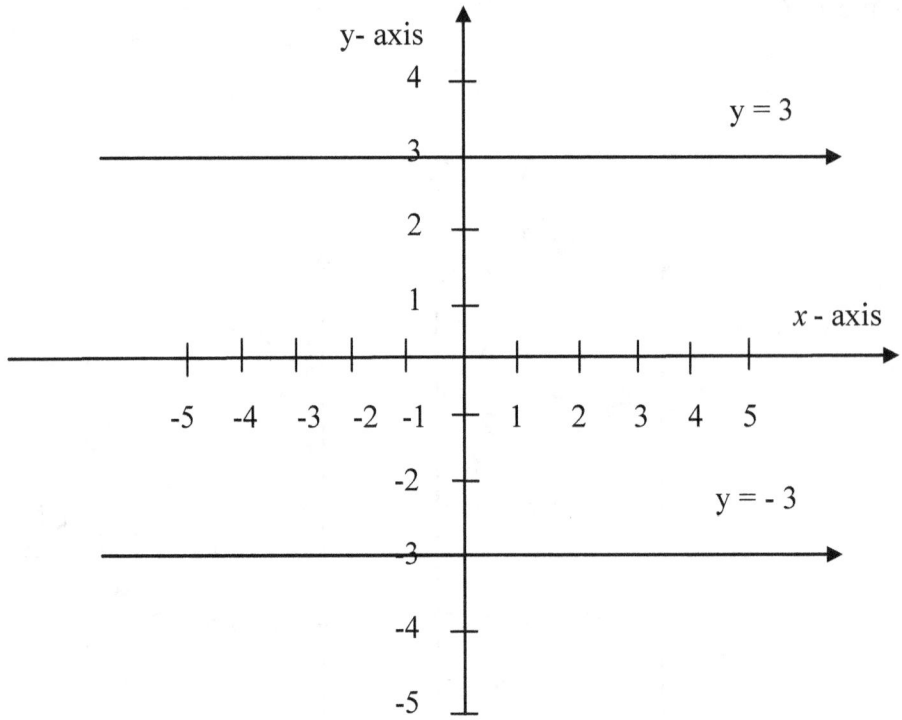

Figure 10.7

A closer look at these special types of straight line equations will reveal the following:

- If the equation is in the form of $x = c$, such as $x = 5$, the x - coordinate (abscissa) of every point on this line is 5.

-   If the equation is in the form $x = -c$, such as $x = -2$,
    the x - coordinate (abscissa) of every point on this line is -2.

-   If the equation is in the form $y = c$, such as $y = 4$,
    the y-coordinate (ordinate) of every point on this line is 4.

-   If the equation is in the form $y = -c$, such as $y = -3$,
    the y-coordinate (ordinate) of every point on this line is -3.

X - intercept and y-intercept

The x - coordinate of the point where a line intersects the x axis is called the x - intercept. For example, if a line intersects the x - axis at the point (3,0), 3 is called the x - intercept.

Similarly, The y-coordinate of the point where a line intersects the y-axis is called the y- intercept. For example, if a line intersects the y axis at the point (0,2), 2 is called the y- intercept. See figure 10.8

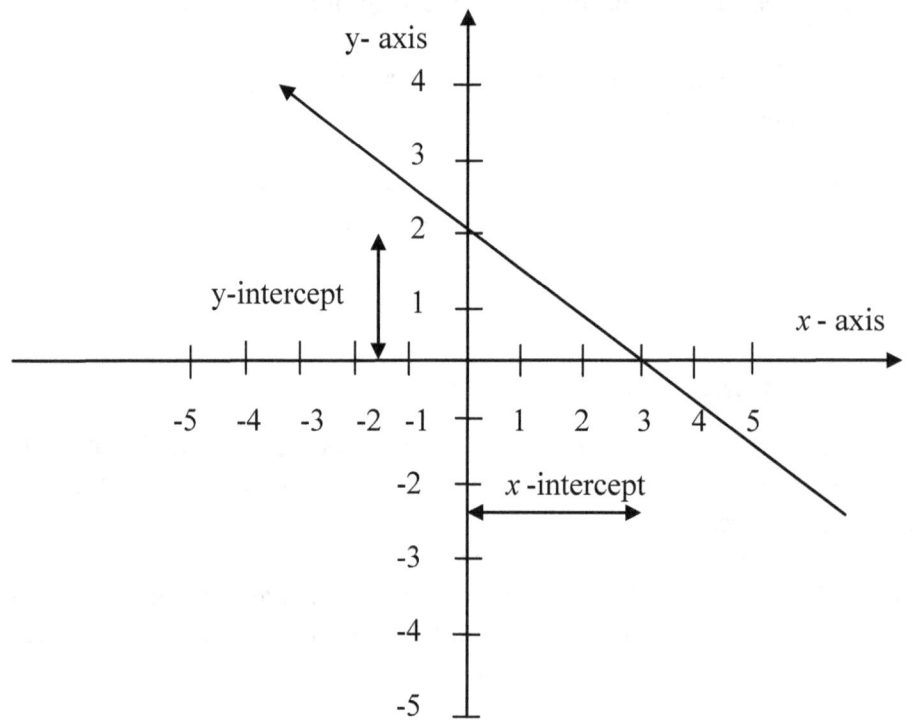

Figure 10.8

Slope of a straight line

The slope of a straight line can be calculated from the coordinates of two different points on the line. To calculate the slope of a line subtract the two y-coordinates of the two points, subtract the two x-coordinates of the two points, then divide the difference between the two y-coordinates by the difference between the two x-coordinates. The resulting quotient is the slope of the line.

172

Slope of a straight line .... $= \dfrac{y_2 - y_1}{x_2 - x_1}$

Where x_2 does not equal x_1

Example 1: Calculate the slope of the straight line that passes through the points (2, 3), and (4, 6).

$y_2 - y_1 = 6 - 3 = 3$
$x_2 - x_1 = 4 - 2 = 2$

Slope of the line $= \dfrac{3}{2}$

Example 2: Calculate the slope of the straight line that passes through the points $(1, -2)$, and $(3, -6)$.

$y_2 - y_1 = -6 - (-2) = -6 + 2 = -4$
$x_2 - x_1 = 3 - 1 = 2$

Slope of the line $= \dfrac{-4}{2} = -2$

The slope of a straight line can also be calculated from the straight line equation.
If the straight line equation is in the form $y = mx + b$
The slope of the line is m (the coefficient of x), and b is the y-intercept of the line.

Example 1: Calculate the slope of the straight line $2y - 6x = 4$

Rewrite the equation in the form $y = mx + b$
$2y = 6x + 4$
Divide each side of the equation by 2
$y = 3x + 2$

The slope of the line is 3 (the coefficient of x).

Example 2:   Calculate the slope of the straight line $2x + 5y = 3$

Rewrite the equation in the form $y = mx + b$
$5y = -2x + 3$
Divide each side of the equation by 5

$$y = \frac{-2}{5}x + \frac{3}{5}$$

The slope of the line is $\dfrac{-2}{5}$ (the coefficient of x).

Special slopes

As explained in this chapter the slope of a straight line $= \dfrac{y_2 - y_1}{x_2 - x_1}$

If $y_2 = y_1$, the numerator of this fraction would equal 0, and the resulting quotient $= 0$. This means that the slope of this particular line is 0, and it indicates that the line is a horizontal line parallel to the x - axis

If $x_2 - x_1 = 0$, the slope of this particular line is undefined, and the line is a vertical line parallel to the y-axis.

When a line intersects another line at an angle of 90 degrees, the two lines are said to be perpendicular to each other. The product of the slopes of two perpendicular lines is equal to -1 , and is denoted by $m_1 m_2 = -1$
Where m_1 is the slope of the first line, and m_2 is the slope of the second line.

If two lines have the same slope, then the two lines are parallel to each other.

Equations of a straight line

Point slope form

The equation of a straight line that passes through a point whose coordinates are (x_1, y_1), with a slope m is ………………………… $y - y_1 = m(x - x_1)$

Example 1: Find the equation of a straight line that passes through the point $(1, 5)$, and a slope of 2.

Substitute 1 for x_1, and 5 for y_1 in the point slope form, and 2 for the slope m.

$$y - 5 = 2(x - 1)$$
$$y - 5 = 2x - 2$$

Add 5 to each side of the equation
$$y - 5 + 5 = 2x - 2 + 5$$
$$y = 2x + 3$$

Example 2: Find the equation of a straight line that passes through the point $(-2, -3)$, and a slope of -1.

Substitute -2 for x_1, and -3 for y_1 in the point slope form, and -1 for the slope m.

$$y - (-3) = -1[(x - (-2)]$$
$$y + 3 = -1(x + 2)$$
$$y + 3 = -x - 2$$

Add -3 to each side of the equation
$$y + 3 - 3 = -x - 2 - 3$$
$$y = -x - 5$$

Equation of a straight line using two points

To find the equation of a straight line using the coordinates of two points on the line, use the following equation

$$\frac{y - y_1}{x - x_1} = \frac{y_2 - y_1}{x_2 - x_1}$$

Example 1:   Find the equation of a straight line that passes through the
               two points (2, 4), and (1, – 2).

$$\frac{y - y_1}{x - x_1} = \frac{y_2 - y_1}{x_2 - x_1}$$

$$\frac{y - 4}{x - 2} = \frac{-2 - 4}{1 - 2}$$

$$\frac{y - 4}{x - 2} = \frac{-6}{-1} = 6$$

$$y - 4 = 6(x - 2)$$
$$y - 4 = 6x - 12$$

Add 4 to each side of the equation

$$y - 4 + 4 = 6x - 12 + 4$$
$$y = 6x - 8$$

Example 2:   Find the equation of a straight line that passes through the
               two points (– 2, – 3), and (3, 7).

$$\frac{y - y_1}{x - x_1} = \frac{y_2 - y_1}{x_2 - x_1}$$

$$\frac{y - (-3)}{x - (-2)} = \frac{7 - (-3)}{3 - (-2)}$$

$$\frac{y + 3}{x + 2} = \frac{10}{5} = 2$$

$$y + 3 = 2(x + 2)$$
$$y + 3 = 2x + 4$$

Add – 3 to each side of the equation

$$y + 3 - 3 = 2x + 4 - 3$$
$$y = 2x + 1$$

Exercise 10.1

Show the location of the following points:

1-	$(2, 4)$	2-	$(3, -3)$	3-	$(4, 0)$
4-	$(-2, 3)$	5-	$(-3, 0)$	6-	$(-1, -4)$
7-	$(0, 4)$	8-	$(-4, 2)$	9-	$(0, -3)$
10-	$(5, -2)$	11-	$(-3, 3)$	12-	$(3, 3)$
13-	$(1, -5)$	14-	$(-2, -2)$		

Exercise 10.2

Graph the following straight lines:

1-	$x - y = 0$	2-	$x + y = 2$
3-	$x - y = 3$	4-	$2x - y = 4$
5-	$2x + y = 6$	6-	$2x + 2y = 0$
7-	$x - 2y = 2$	8-	$x + 2y = 4$
9-	$2x - 2y = 6$	10-	$3x + y = 12$
11-	$3x + 2y = 12$	12-	$3x - 2y = 6$
13-	$x = -2$	14-	$x = 5$
15-	$y = 3$	16-	$y = -4$
17-	$2x + 3y = 0$	18-	$2x - 3y = 12$
19-	$8x - 4y = 8$	20-	$5x + 2y = 10$
21-	$4y - x = 8$	22-	$y + 2x = -4$
23-	$y - 2x = 4$	24-	$3x + 3y = 9$

Exercise 10.3

Find the slopes of the following straight lines:

1-	$y - x = 0$	2-	$y + x = 0$
3-	$2y - x = 0$	4-	$x - 2y = 0$
5-	$x + 2y = 0$	6-	$x - 3y = 3$
7-	$2x - y = 4$	8-	$2x + 2y = 6$
9-	$2x - 3y = 12$	10-	$y - 2x = 5$
11-	$2y - 3x = 12$	12-	$x - y = 5$
13-	$2y - 2x = 8$	14-	$3y - x = 9$
15-	$3y - 3x = 3$	16-	$3y - 4x = 12$
17-	$3y + x = 6$	18-	$3y + 3x = 9$
19-	$x + 4y = 8$	20-	$x - 5y = 10$

Exercise 10.4

Find the equations of the straight lines, given the following information:

A- Point – slope problems

1-	$(2, 1); 1$	2-	$(3, 2); 3$
3-	$(-1, 4); -2$	4-	$(2, 3); -3$
5-	$(-2, -3); -4$	6-	$(3, -4); 2$
7-	$(-3, 4); \dfrac{3}{2}$	8-	$(1, -5); 1$
9-	$(-4, -1); -3$	10-	$(-2, -4); 4$

B- Two points problems

11-	$(0, 0); (2, 2)$	12-	$(0, 1); (4, 4)$
13-	$(1, 2); (3, 5)$	14-	$(-1, 2); (2, 3)$
15-	$(-2, 4); (1, 4)$	16-	$(-3, -1); (-1, 1)$
17-	$(-4, -2); (3, 3)$	18-	$(-1, -1); (3, 4)$
19-	$(-2, 3); (1, -3)$	20-	$(-1, 4); (1, -4)$

CHAPTER 11

TWO LINEAR EQUATIONS AND INEQUALITIES

Solving two linear equations in two variables

Substitution method

To find the solution of two linear equations in two variables use the following steps:

➢ Start with the first equation and find the value of one of the two variables in terms of the other variable.

➢ Substitute this value for the variable in the second equation; this should result in an equation in one variable.

➢ Solve the second equation, and find the value of the variable

➢ Substitute the value obtained for this variable in the first equation and calculate the value of the other variable.

Example 1: Solve the following two linear equations:
$x - y = 4$; and $2x + y = 11$

Find the value of x in terms of y from the first equation.
$x - y = 4$
$x = y + 4$

Substitute this value of x in the second equation.
$2x + y = 11$
$2(y + 4) + y = 11$
$2y + 8 + y = 11$

$3y = 3$
$y = 1$

Substitute the value of y in the first equation and calculate the value of x.
$x - y = 4$
$x - 1 = 4$
$x = 5$

The solution is (5, 1)

Example 2: Solve the following two linear equations:
$x + 3y = 1$; and $2x - y = 9$

Find the value of x in terms of y from the first equation.
$x + 3y = 1$
$x = 1 - 3y$

Substitute this value of x in the second equation.
$2x - y = 9$
$2(1 - 3y) - y = 9$
$2 - 6y - y = 9$
$2 - 7y = 9$
$-7y = 7$
$y = -1$

Substitute the value of y in the first equation and calculate the value of x.
$x + 3y = 1$
$x - 3 = 1$
$x = 4$

The solution is (4, − 1)

Elimination method

To find the solution of two linear equations in two variables use the following steps:

> Compare the values of one of the variables in the two equations.

> Try to eliminate one of the variables by choosing a number (positive or negative) such that if multiplied by the first equation, and (if necessary) another number (positive or negative) such that if multiplied by the second equation will result in the same coefficient with opposite signs.

> Multiply each number explained above by the corresponding equation, and add the two new equations.

> Calculate the value of the remaining variable.

> Substitute the value obtained for this variable in one of the original equations and calculate the value of the other variable.

Example 1:   Solve the following two linear equations:
$$x + 2y = -4; \text{ and } 2x + 3y = -5$$

Comparing the values of the x terms in the two equations, we find out that if we multiply the first equation by -2, it will yield an equation such that if the new equation is added to the second equation the term in x will be eliminated.

Multiply the first equation by -2
$$-2(x + 2y) = -2(-4)$$
$$-2x - 4y = 8$$

Add this equation to the second equation

$$-2x - 4y = 8$$
$$2x + 3y = -5$$

$$-y = 3 \quad \text{or} \quad y = -3$$

Substitute -3 for the value of y in the first equation
$$x + 2(-3) = -4$$
$$x - 6 = -4$$
$$x = 2$$

The solution is $(2, -3)$

Example 2:   Solve the following two linear equations:
$$4x - 2y = 14; \text{ and } 3x + 3y = 24$$

Comparing the values of the x terms in the two equations, we find out that if we multiply the first equation by 3, and the second equation by -4 it will yield two new equations such that if the new equations are added the term in x will be eliminated.

Multiply the first equation by 3
$$3(4x - 2y) = 3(14)$$
$$12x - 6y = 42$$

Multiply the second equation by -4
$$-4(3x + 3y) = -4(24)$$
$$-12x - 12y = -96$$

Add the two new equations
$$12x - 6y = 42$$
$$-12x - 12y = -96$$

$$-18y = -54 \quad \text{or } y = 3$$

Substitute 3 for the value of y in the first equation

$4x - 2(3) = 14$

$4x - 6 = 14$

$4x = 20$ or $x = 5$

The solution is (5, 3)

Graphical method

To find the solution set of two linear equations in two variables, graph the two equations. The coordinates of the point of intersection of the two lines is the solution set of the two equations. If the two lines are parallel to each other, or if the equations have no solution, the graph will not yield a point of intersection.

Example: Solve the following two linear equations graphically:

$x - 2y = -2$; and $2x - y = 5$

Graph the two equations using the method explained in chapter 10 for graphing linear equations.

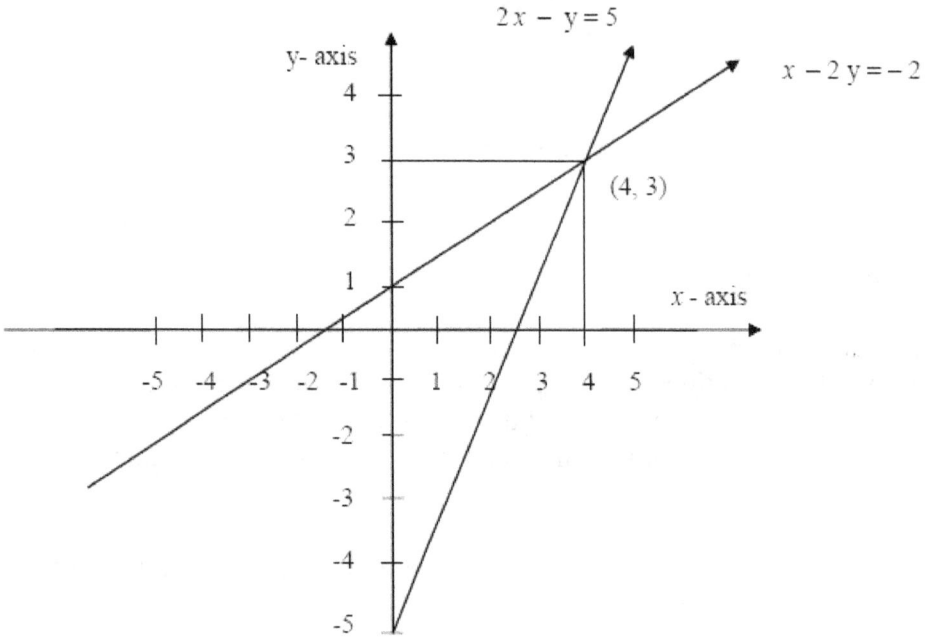

The solution is (4, 3)

Graphing linear inequalities in two variables

Graphing linear equations was explained in chapter 10. The methods used to graph a straight line will be used to graph linear inequalities in two variables. The solution of linear inequalities in two variables such as $x + y > 3$ includes an infinite number of points, usually shown as a shaded area, above or below the line represented by the equation $x + y = 3$. The shaded area also includes the straight line itself if the inequality is in the form of $x + y \geq 3$ or $x + y \leq 3$

To graph a linear inequality in two variables such as $x + y > 3$ do the following:

➤ Graph the line represented by the equation $x + y = 3$

➤ The solution set of the linear inequalities $x + y > 3$ and $x + y < 3$ will include either the area above the straight line or the area below the straight line. In these two cases the line is shown as a dashed line in the graph to indicate that the line itself is not part of the solution.

If the inequality is in the form of $x + y \geq 3$ or $x + y \leq 3$, the solution set will also include the straight line itself in addition to the area above or below the straight line. In these two cases the line is shown as a solid line in the graph to indicate that the line itself is part of the solution.

➤ To determine whether the solution points are located above or below the line choose the coordinates of any point above or below the line (but not on the line), and substitute the values chosen for x and y in the linear inequality. If the result is a true statement that satisfies the inequality, then the side where the chosen point is located is the side to be shaded.

If the resulting statement is not true, then the solution set of the inequality will be on other side of the straight line, and this will be the area to shade.

Example 1: Graph the inequality $x + y > 3$

Graph the straight line $x + y = 3$
Choose the coordinates of a point (any point) such as (0, 0).
Substitute 0 for x and 0 for y in the inequality
The result is $0 + 0 > 3$, or $0 > 3$.

Since this is not a true statement, then the solution set of the inequality will be on the other side of the straight line (the side opposite to where the point (0,0) is located), and this will be the area to shade. See figure 11.1
Note that the line is shown as a dashed line in the graph to indicate that the line itself is not part of the solution.

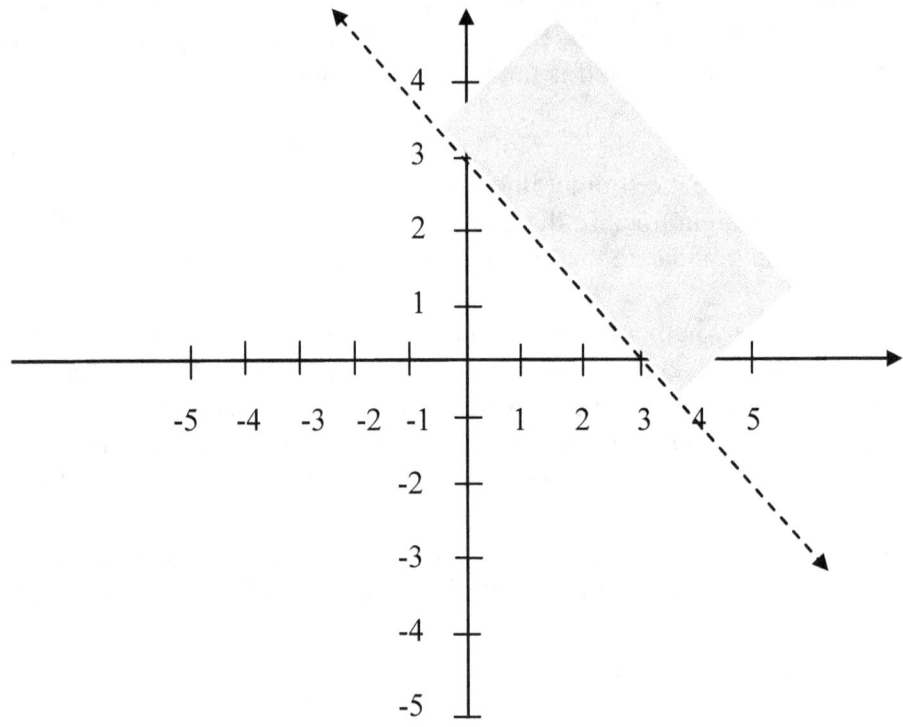

Figure 11.1

Example 2:   Graph the inequality $x - y < -3$

Graph the straight line $x - y = -3$
Choose the coordinates of a point (any point) such as (0, 0).
Substitute 0 for x and 0 for y in the inequality
The result is $0 - 0 < 3$, or $0 < -3$.

186

Since this is not a true statement, then the solution set of the inequality will be on the other side of the straight line (the side opposite to where the point (0,0) is located), and this will be the area to shade. See figure 11.2

Note that the line is shown as a dashed line in the graph to indicate that the line itself is not part of the solution.

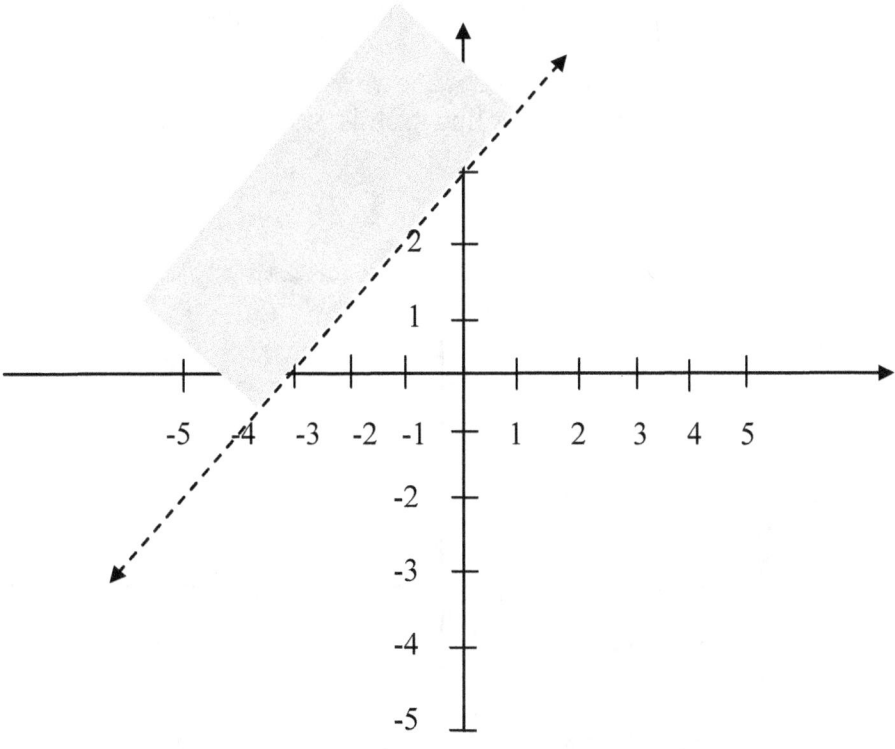

Figure 11.2

Example 3:   Graph the inequality $-2\,y - x \geq 4$

Graph the straight line $-2\,y - x = 4$
Choose the coordinates of a point (any point) such as (0, 0).
Substitute 0 for x and 0 for y in the inequality
The result is $0 - 0 \geq 4$, or $0 \geq 4$.

Since this is not a true statement, then the solution set of
the inequality will be on the other side of the straight line
(the side opposite to where the point (0,0) is located), and
this will be the area to shade. See figure 11.3

Note that the line is shown as a solid line in the graph to
indicate that the line itself is part of the solution.

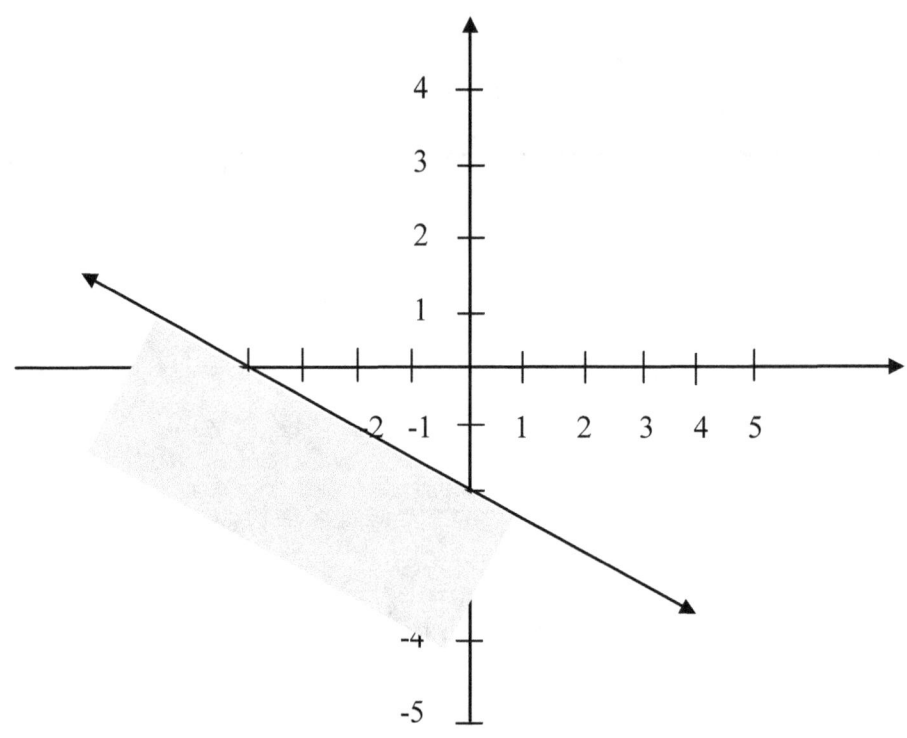

Figure 11.3

188

Example 4:   Graph the inequality $y - 2x \leq -4$

Graph the straight line $y - 2x = -4$
Choose the coordinates of a point (any point) such as (0, 0).
Substitute 0 for x and 0 for y in the inequality
The result is $0 - 0 \leq -4$, or $0 \leq -4$.

Since this is not a true statement, then the solution set of
the inequality will be on the other side of the straight line
(the side opposite to where the point (0,0) is located), and
this will be the area to shade. See figure 11.4

Note that the line is shown as a solid line in the graph to
indicate that the line itself is part of the solution.

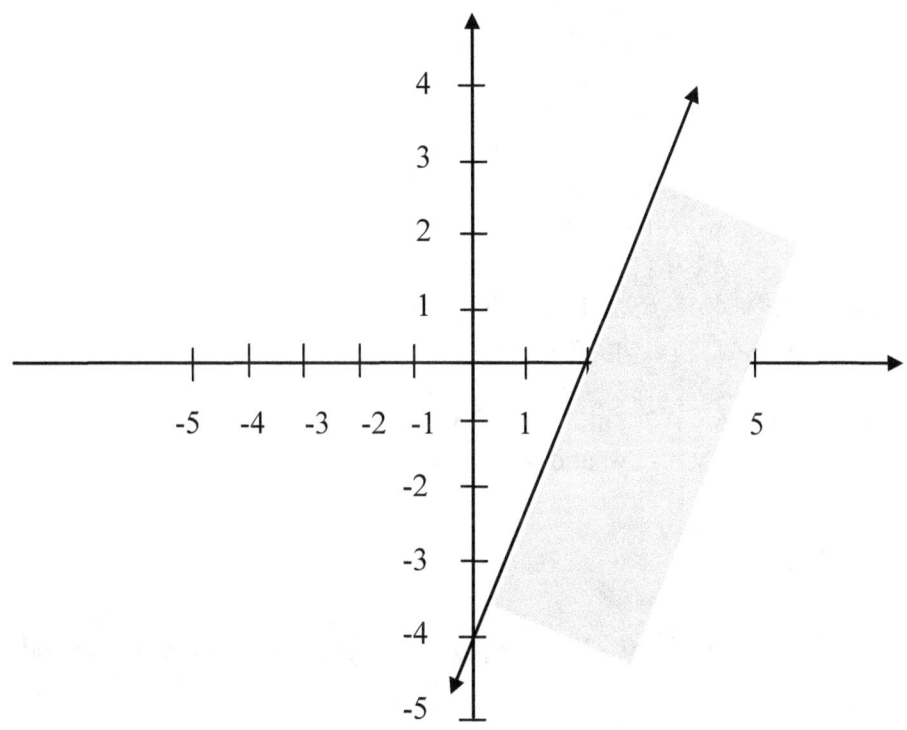

Figure 11.4

189

Exercise 11.1

Solve the following two linear equations using the substitution method:

1- $x - y = 0$; and $2x + 3y = 15$
2- $x + 2y = 0$; and $2x + 5y = -2$
3- $x - 3y = 0$; and $2x - y = 10$
4- $2x - y = 1$; and $3x + 2y = 12$
5- $2x + y = 11$; and $3x - 2y = -1$
6- $2x - 2y = -6$; and $2x + 3y = 14$
7- $2x - 3y = -15$; and $x + 4y = 20$
8- $3x - y = 2$; and $x - 3y = -10$
9- $3x + y = 12$; and $x + 3y = 4$
10- $3x + 2y = 3$; and $2x + 4y = 10$
11- $3x + 4y = -7$; and $4x - y = -3$
12- $4x + y = -2$; and $x + 2y = 3$
13- $4x - 3y = 9$; and $2x - 2y = 6$
14- $4x - y = -1$; and $2x + 3y = 17$
15- $4x + 2y = 14$; and $3x + 3y = 12$
16- $5x - y = 18$; and $4x + 2y = 20$
17- $5x - 2y = 25$; and $3x + y = 15$
18- $3x + 4y = 19$; and $2x - 2y = 8$
19- $3x + 5y = 17$; and $4x - 2y = 14$
20- $2x + 4y = 6$; and $3x + 3y = 9$
21- $3x - y = 13$; and $4x + 4y = 12$
22- $4x - 3y = 7$; and $5x - y = 17$
23- $2x + 6y = 20$; and $5x - 2y = -1$
24- $3x - 4y = -24$; and $4x + 3y = 18$

Exercise 11.2

Solve the following two linear equations using the elimination method:

1- $x - 2y = 0$; and $x + 3y = 15$
2- $x - 3y = 0$; and $x + 4y = 14$
3- $x - 4y = 0$; and $x + 5y = 18$

4- $x + y = 3$; and $2x + 4y = 10$

5- $x + 2y = 8$; and $2x + y = 4$

6- $2x + 2y = 14$; and $3x + y = 13$

7- $2x + 3y = 12$; and $4x + y = 14$

8- $3x - y = 4$; and $4x - y = 6$

9- $3x + y = 16$; and $4x - 2y = 8$

10- $4x + 2y = 24$; and $x - y = 3$

11- $6x + y = 23$; and $2x - 2y = -4$

12- $2x - 3y = -13$; and $5x - 2y = -5$

13- $2x + 4y = 16$; and $6x - y = -4$

14- $5x - 3y = 17$; and $x + 5y = 9$

15- $2x - 5y = 7$; and $4x - 5y = 19$

16- $3x + 3y = 21$; and $5x - 5y = -15$

17- $4x + 3y = 13$; and $5x - 2y = -1$

18- $7x - y = 10$; and $2x + 3y = 16$

19- $2x - 3y = 7$; and $5x - 4y = 21$

20- $6x - 2y = 16$; and $x + 5y = 8$

21- $7x - 2y = 15$; and $3x + 4y = 21$

22- $4x + 2y = 26$; and $2x - 3y = -7$

23- $3x + 2y = 22$; and $5x - 10y = 10$

24- $4x + y = 10$; and $3x - 2y = -9$

Exercise 11.3

Using the graphical method find the solution of the following equations:

1- $x = 3$; and $x + y = 5$

2- $x = -2$; and $x + 2y = 6$

3- $y = -3$; and $x - 3y = 5$

4- $y = 3$; and $x + 4y = 9$

5- $x + y = 4$; and $x - 2y = 7$

6- $x + y = 0$; and $2x + y = 3$

7- $x - y = 5$; and $3x + y = 15$

8- $x - y = 6$; and $2x - 3y = 11$

9- $2x + 3y = 19$; and $3x - 2y = -4$

10- $x - 2y = -7$; and $2x + y = 6$
11- $3x - 2y = 3$; and $2x + 3y = 2$
12- $2x - y = 8$; and $x + 2y = 4$
13- $x + 2y = -6$; and $2x - 2y = -4$
14- $2x + 2y = -2$; and $3x - 3y = -15$
15- $4x + y = 5$; and $x + 3y = -7$
16- $3x - y = -10$; and $x + 4y = -12$
17- $x = y$; and $2x + 5y = 7$
18- $x - 3y = 0$; and $3x - 4y = 5$
19- $3x + y = -2$; and $2x + 3y = 8$
20- $x - 4y = 4$; and $4x - 5y = -6$
21- $5x + y = 3$; and $3x + 2y = 6$
22- $4x - 3y = -7$; and $2x + y = -1$
23- $3x - 3y = -9$; and $2x + 2y = -6$
24- $3x + 3y = -9$; and $2x - y = 5$

Exercise 11.4

Graph the following inequalities:

1- $x > 2$ 2- $x < -3$
3- $x \geq 4$ 4- $x \leq -4$
5- $y > 2$ 6- $y < -3$
7- $y \geq 4$ 8- $y \leq -4$
9- $x - y < 0$ 10- $x + y < 0$
11- $x + y \geq 2$ 12- $x - 2y \leq 2$
13- $x + 2y > 2$ 14- $x - y \leq 3$
15- $x - y \geq 3$ 16- $2x + y > 0$
17- $2x + y < 3$ 18- $2x - y \geq 4$
19- $3x - y < 2$ 20- $3x + y > 1$
21- $x + 3y > 3$ 22- $x - 3y > 2$
23- $2x + 2y < 4$ 24- $2x - 2y \leq 2$

CHAPTER 12

GRAPHING QUADRATIC EQUATIONS

The graph of a quadratic equation is a symmetrical curved shape called a **parabola**.

Parabolas that are open at the top or the bottom

The general form of quadratic equations that result in a parabola open at the top or the bottom is …….. $y = ax^2 + bx + c$
Where a, b, and c are constants, and a does not equal zero.

The main elements of a parabola are the vertex and the axis of symmetry. The vertex is the highest point in the parabolas that are open at the bottom, and the vertex is the lowest point in the parabolas that are open at the top.

Parabolas that are open at the top or the bottom are symmetrical about a vertical line that passes through the vertex and the line is called the **axis of symmetry**. See Figure 12.1.

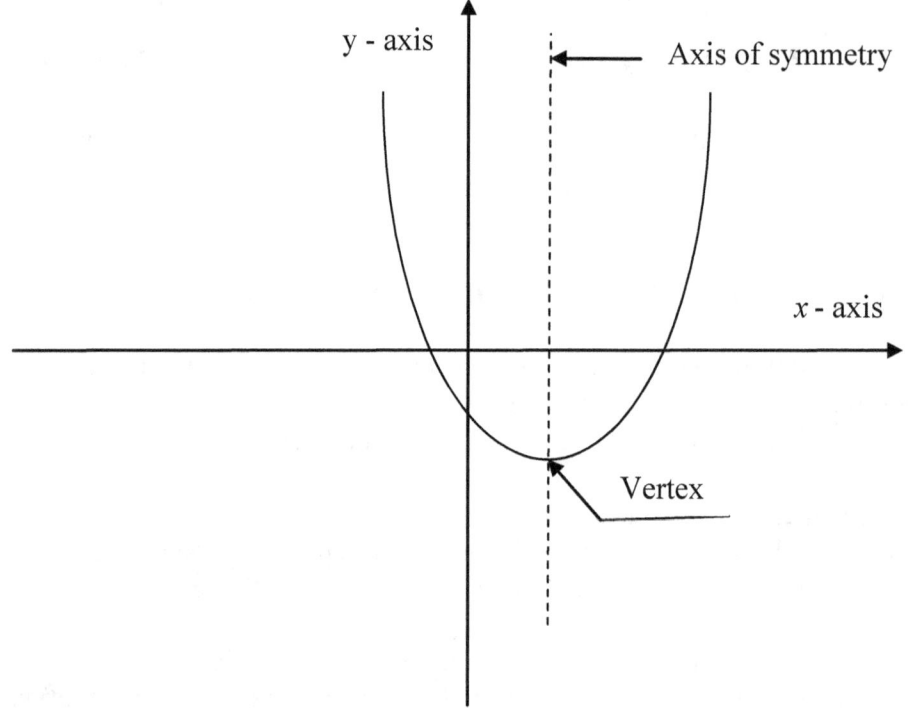

Figure 12.1

The vertex of a parabola

There are two methods to find the coordinates of the vertex.

Method 1:

If the quadratic equation is of the form $y = a x^2 + b x + c$
The x - coordinate of the vertex can be calculated using the following formula:

$$x = \frac{-b}{2a}$$

To find the y- coordinate of the vertex, substitute the calculated x value in the quadratic equation, and find the value of y.

Method 2:

If the quadratic equation is of the form, or can be converted to the form, $y = a (x - h)^2 + k$, where "a" does not equal zero, then the coordinates of the vertex are (h, k).

The axis of Symmetry of a parabola

The axis of symmetry is a vertical line that passes through the vertex, and the parabola is symmetrical about this line. There are two methods to find the equation of the axis of symmetry.

Method 1:

If the quadratic equation is of the form $y = a x^2 + b x + c$

the equation of the axis of symmetry is ..... $x = \dfrac{-b}{2a}$

Method 2:

If the quadratic equation is of the form, or can be converted to the form, $y = a (x - h)^2 + k$, where a does not equal zero, then the equation of the axis of symmetry is ..... $x = h$

Is the parabola open at the top or the bottom?

To determine whether a parabola is open at the top or the bottom, look at the value of the constant a:

If a > 0, the parabola is open at the top.
If a < 0, the parabola is open at the bottom.

Points of intersection with the x-axis

The graph of a parabola can be totally above the x-axis, below the x-axis, or intersect with the x-axis. For the parabolas that intersect with the x-axis, to find the coordinates of the points of intersection with the x-axis do the following:

- ➤ Substitute 0 for the value of y in the equation.

- ➤ Solve the equation and find the two values of x (let's call them x_1, and x_2)

- ➤ The coordinates of the points of intersection with the x-axis are $(x_1, 0)$, and $(x_2, 0)$

Graphing parabolas that are open at the top or the bottom

To graph a parabola we need the coordinates of at least three points on the parabola, the vertex can be one of these three points.

To find the coordinates of points on the parabola, assume values for x, substitute each value in the quadratic equation, and find the corresponding values of y.

Example 1:   Graph the following parabola:
$$y = x^2 - 2$$

Note that the term (b x) is missing from this equation.
a = 1   b = 0   c = -2

The x-coordinate of the vertex is:
$$x = \frac{-b}{2a} = \frac{-0}{2(1)} = 0$$

Substitute 0 for x in the equation and calculate the value of y.

196

$y = (0)^2 - 2$ or $y = -2$
The coordinates of the vertex are $(0, -2)$.

Assume values for x, substitute these values in the
equation, and find the corresponding values of y.

x	1	-1	2	-2
y	-1	-1	2	2

Plot these points and connect with a curve. See figure 12.2.

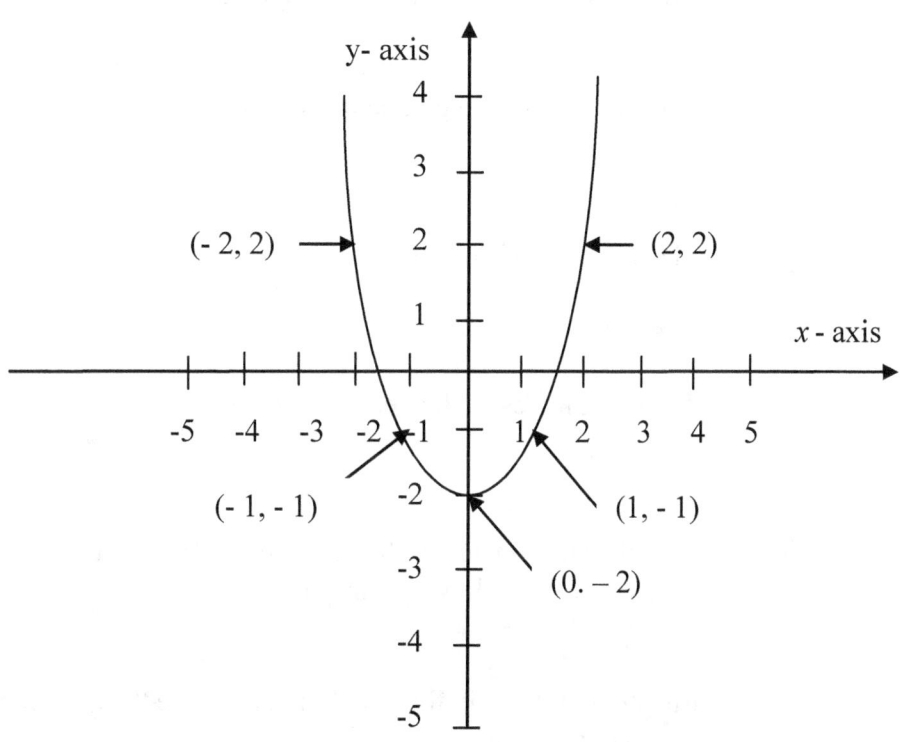

Figure 12.2

197

Example 2: Find the equation of the axis of symmetry of the following parabola:

$$y = x^2 - 4x + 4$$

$a = 1$ $b = -4$ $c = 4$

The equation of the axis of symmetry is

$$x = \frac{-b}{2a}$$

$$x = \frac{-(-4)}{2(1)} = \frac{4}{2} = 2$$

The equation of the axis of symmetry is $x = 2$

Example 3: Find the coordinates of the vertex, and the equation of the axis of symmetry of the following parabola:

$$y = (x + 2)^2$$

The equation of the parabola is of the form

$$y = a(x - h)^2 + k$$

Where $a = 1$ $h = -2$ $k = 0$

The equation of the axis of symmetry is

$$x = h$$
$$x = -2$$

The coordinates of the vertex are (h, k)
$$= (-2, 0)$$

Example 4: Find the coordinates of the points of intersection with the x - axis of the following parabola:

$$y = x^2 - 4$$

Substitute 0 for y in the equation, and solve the equation to find the two values of x.

$$x^2 - 4 = 0$$
$$x^2 = 4$$

Take the square root of each side of the equation

$x = 2$, or $x = -2$

The coordinates of the points of intersection with the x-axis are $(2, 0)$, and $(-2, 0)$

Parabolas that are open to the right or to the left

The general form of quadratic equations that result in a parabola that is open to the right or to the left is…….. $x = ay^2 + by + c$
Where a, b, and c are constants, and a does not equal zero.

The main elements of a parabola are the vertex and the axis of symmetry. The vertex is the farthest point to the right in the parabolas that are open to the left, and the vertex is the farthest point to the left in the parabolas that are open to the right.

Parabolas that are open to the right or to the left are symmetrical about a horizontal line that passes through the vertex and the line is called the axis of symmetry. See Figure 12.3.

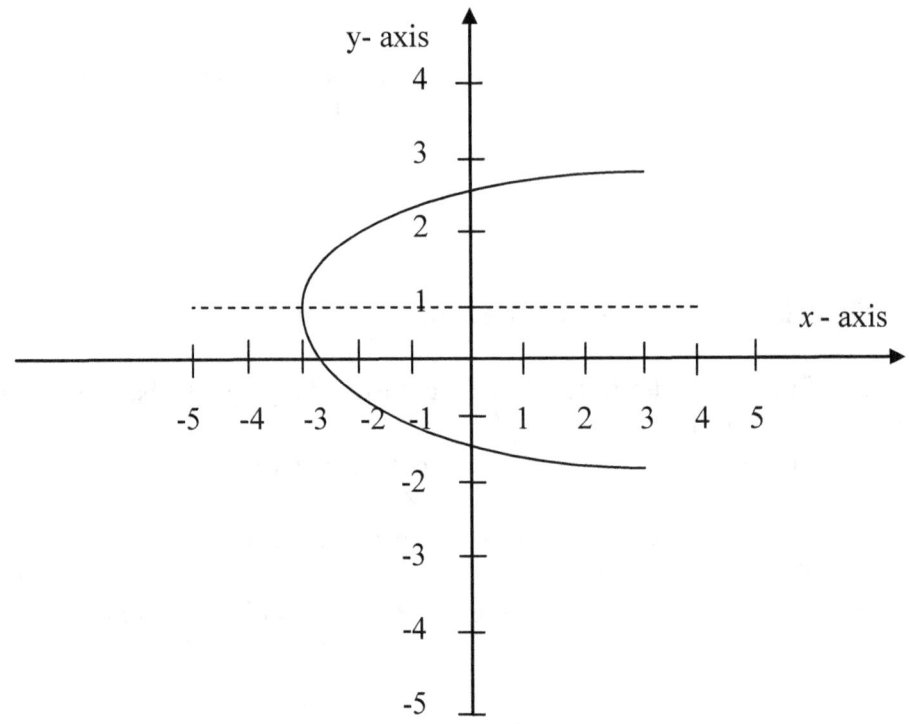

Figure 12.3

The vertex of a parabola

There are two methods to find the coordinates of the vertex.

Method 1:

If the quadratic equation is of the form $x = ay^2 + by + c$
The y- coordinate of the vertex can be calculated using the following formula:
$$y = \frac{-b}{2a}$$
To find the x - coordinate of the vertex, substitute the calculated y value in the quadratic equation, and find the value of x.

200

Method 2:

If the quadratic equation is of the form, or can be converted to the form, $x = a(y-h)^2 + k$, where a does not equal zero, then the coordinates of the vertex are (h, k).

The axis of Symmetry of a parabola

The axis of symmetry is a vertical line that passes through the vertex, and the parabola is symmetrical about this line. There are two methods to find the equation of the axis of symmetry.

Method 1:

If the quadratic equation is of the form $x = ay^2 + by + c$

the equation of the axis of symmetry is ..... $y = \dfrac{-b}{2a}$

Method 2:

If the quadratic equation is of the form, or can be converted to the form, $x = a(y-h)^2 + k$, where a does not equal zero, then
the equation of the axis of symmetry is ..... $y = h$

Is the parabola open to the right or to the left?

To determine whether a parabola is open to the right or to the left, look at the value of the constant a:

If a > 0, the parabola is open to the right.
If a < 0, the parabola is open to the left.

Points of intersection with the y-axis

The graph of a parabola can be totally to the left of the y-axis, to the right of the y-axis, or intersect with the y-axis. For the parabolas that intersect with the y-axis, to find the coordinates of the points of intersection with the y-axis do the following:

➢ Substitute 0 for the value of x in the equation.

➢ Solve the equation and find the two values of y (let's call them y_1, and y_2)

➢ The coordinates of the points of intersection with the y-axis are $(0, y_1)$, and $(0, y_2)$

Graphing parabolas that are open to the right or to the left

To graph a parabola we need the coordinates of at least three points on the parabola, the vertex can be one of these three points.

To find the coordinates of points on the parabola, assume values for y, substitute each value in the quadratic equation, and find the corresponding values of x.

Example 1: Graph the following parabola:
$$x = y^2 - 2$$

Note that the term (b y) is missing from this equation.
$$a = 1 \qquad b = 0 \qquad c = -2$$

The y- coordinate of the vertex is:
$$y = \frac{-b}{2a} = \frac{-0}{2(1)} = 0$$

Substitute 0 for y in the equation and calculate the value of x.

$x = (0)^2 - 2$ or $x = -2$

The coordinates of the vertex are $(-2, 0)$.

Assume values for y, substitute these values in the equation, and find the corresponding values of x.

y	1	-1	2	-2
x	-1	-1	2	2

Plot these points and connect with a curve.
See figure 12.4

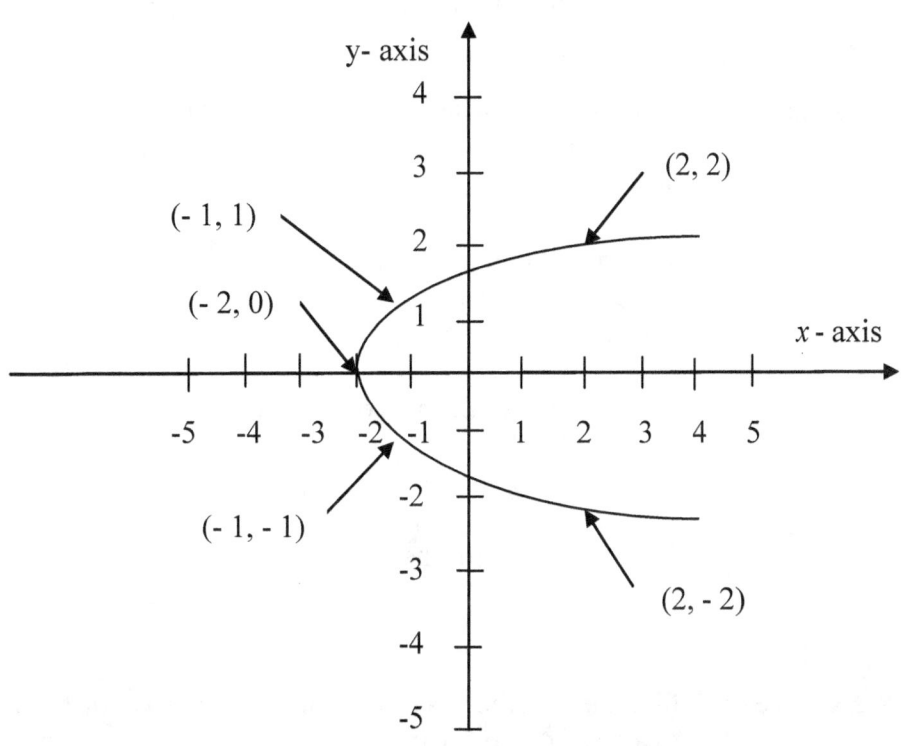

Figure 12.4

203

Example 2:   Find the equation of the axis of symmetry of the following parabola:

$$x = y^2 - 4y + 4$$

$a = 1$ $b = -4$ $c = 4$

The equation of the axis of symmetry is

$$y = \frac{-b}{2a}$$

$$y = \frac{-(-4)}{2(1)} = \frac{4}{2} = 2$$

The equation of the axis of symmetry is $y = 2$

Example 3:   Find the coordinates of the vertex, and the equation of the axis of symmetry of the following parabola:

$$x = (y + 2)^2$$

The equation of the parabola is of the form
$$x = a(y - h)^2 + k$$
Where $a = 1$ $h = -2$ $k = 0$

The equation of the axis of symmetry is
$y = h$
$y = -2$

The coordinates of the vertex are (h, k)
$= (-2, 0)$

Example 4:   Find the coordinates of the points of intersection with the y- axis of the following parabola:

$$x = y^2 - 4$$

Substitute 0 for x in the equation, and solve the equation to find the two values of y.

$$y^2 - 4 = 0$$
$$y^2 = 4$$

Take the square root of each side of the equation

$$y = 2 \ \text{ or } \ y = -2$$

The coordinates of the points of intersection with the y- axis are $(0, 2)$, and $(0, -2)$

Exercise 12.1

Indicate whether the following parabolas are open at the top or the bottom:

1-	$y = 3x^2$		2-	$y = 5x^2$
3-	$y = -2x^2$		4-	$y = -x^2$
5-	$y = x^2 + 2x$		6-	$y = 3x - x^2$
7-	$y = 4x^2 - 8x + 6$		8-	$y = -2x^2 + 12x - 14$
9-	$y = -3x^2 - 6x + 10$		10-	$y = x^2 + 4x + 8$

Exercise 12.2

Indicate whether the following parabolas are open to the right or to the left:

1-	$x = 4y^2$		2-	$x = -9y^2$
3-	$x = 12y^2$		4-	$x = -y^2$
5-	$x = 2y^2 + 3$		6-	$x = -3y^2 - 6$
7-	$x = 4y^2 + 2y - 1$		8-	$x = -y^2 + 5y - 7$
9-	$x = 5y^2 - 2y - 2$		10-	$x = -6y^2 + 3y + 3$

Exercise 12.3

Find the coordinates of the vertex, and the equation of the axis of symmetry of the following parabolas:

1- $y = 4x^2$ 2- $y = -3x^2$

3- $y = 2x^2 + 4$ 4- $y = 3 - 2x^2$

5- $y = x^2 - 4x + 4$ 6- $y = -2x^2 + 4x - 2$

7- $y = 3x^2 - 18x + 23$ 8- $y = -3x^2 - 3x - 1$

9- $y = 4x^2 + 16x + 20$ 10- $y = -4x^2 - 8x - 6$

11- $x = y^2$ 12- $x = -4y^2$

13- $x = 2y^2 - 5$ 14- $x = -3y^2 + 12$

15- $x = y^2 + 2y + 6$ 16- $x = 5y^2 - 10y - 3$

17- $x = -2y^2 - 6y + 2$ 18- $x = 3y^2 - 6y - 4$

19- $x = -4y^2 + 12y - 10$ 20- $x = 6y^2 + 12y + 6$

Exercise 12.4

Find the coordinates of the points of intersection of the following parabolas with the x - axis:

1- $y = x^2 - 1$ 2- $y = x^2 - 9$

3- $y = x^2 + 3$ 4- $y = 2x^2 + 2$

5- $y = x^2 - 3x$ 6- $y = 3x^2 + 12x$

7- $y = x^2 - 3x + 2$ 8- $y = 2x^2 - 2x - 4$

9- $y = 3x^2 - 18x + 24$ 10- $y = 4x^2 - 8x - 60$

Exercise 12. 5

Find the coordinates of the points of intersection of the following parabolas with the y- axis:

1-	$x = y^2 - 4$	2-	$x = y^2 - 16$
3-	$x = y^2 + 9$	4-	$x = 2y^2 + 4$
5-	$x = 3y^2 - 12$	6-	$x = 2y^2 - 50$
7-	$x = y^2 - 3y + 2$	8-	$x = y^2 - 2y - 3$
9-	$x = y^2 + 2y - 8$	10-	$x = y^2 + 8y + 15$

Exercise 12.6

Graph the following parabolas:

1-	$y = x^2$	2-	$y = -x^2$
3-	$y = x^2 + 2$	4-	$y = 2 - x^2$
5-	$y = x^2 + 4$	6-	$y = 4 - x^2$
7-	$y = x^2 - 3$	8-	$y = -x^2 - 3$
9-	$y = x^2 - 2x + 1$	10-	$y = x^2 + 4x + 4$
11-	$y = -2x^2 + 4x - 1$	12-	$y = 2x^2 - 8x + 11$
13-	$x = y^2$	14-	$x = -y^2$
15-	$x = 4 - y^2$	16-	$x = -y^2 - 2$
17-	$x = 2y^2 - 4y + 3$	18-	$x = -3y^2 - 12y - 8$
19-	$x = 3y^2 + 12y + 9$	20-	$x = -3y^2 - 6y - 5$

APPENDIX

ANSWERS TO ALL EXERCISES

Exercise 1.1

1-	13	2-	-20
3-	13	4-	4
5-	12	6-	8
7-	24	8-	-40
9-	-45	10-	42
11-	-60	12-	72
13-	162	14-	-140
15-	7	16-	4
17-	36	18-	-18
19-	-34	20-	44
21-	-4	22-	10
23-	6	24-	-18
25-	2	26-	3
27-	-6	28-	2
29-	-2	30-	7
31-	-4	32-	1
33-	-2	34-	9

Exercise 1.2

1-	$10 > 8, 10 \not< 8, 10 \neq 8$	2-	$4 < 9, 4 \not> 9, 4 \neq 9$
3-	$3 = 3, 3 \not> 3, 3 \not< 3$	4-	$11 > 6, 11 \not< 6, 11 \neq 6$
5-	$13 < 14, 13 \not> 14, 13 \neq 14$	6-	$16 = 16, 16 \not> 16, 16 \not< 16$
7-	$5 > 3, 5 \not< 3, 5 \neq 3$	8-	$12 = 12, 12 \not> 12, 12 \not< 12$
9-	$0 < 3, 0 \not> 3, 0 \neq 3$	10-	$0 > -4, 0 \not< -4, 0 \neq -4$
11-	$6 > 0, 6 \not< 0, 6 \neq 0$	12-	$-5 < 0, -5 \not> 0, -5 \neq 0$
13-	$0 = 0, 0 \not> 0, 0 \not< 0$	14-	$0 > -7, 0 \not< -7, 0 \neq -7$
15-	$14 > -2, 14 \not< -2, 14 \neq -2$		
16-	$-15 < 4, -15 \not> 4, -15 \neq 4$		
17-	$-7 < -5, -7 \not> -5, -7 \neq -5$		
18-	$12 = 12, 12 \not> 12, 12 \not< 12$		
19-	$-9 > -13, -9 \not< -13, -9 \neq -13$		
20-	$15 > 5, 15 \not< 5, 15 \neq 5$		

Exercise 1.3

1-	$y + x$	2-	$z - a$
3-	ab	4-	c/y
5-	$y < b$	6-	$b > c$
7-	$d = x$	8-	$y \leq a$
9-	$b \geq z$	10-	$\sqrt{16}$
11-	$\sqrt[3]{8}$	12-	$x \pm (y + z)$
13-	25%	14-	$255 \not< 245$
15-	$(x + y) \not> (a + b)$	16-	$\dfrac{b+c}{y+z} \neq \dfrac{a+b}{x+y}$
17-	z^2	18-	b^3
19-	10^4	20-	$\lvert 6 \rvert$
21-	$\lvert -15 \rvert$	22-	$\left\lvert \dfrac{y}{d} \right\rvert$

Exercise 2.1

1- $z + y$
2- Many answers, this is only one of the answers: $6 + 8 - 7$
3- Many answers, this is only one of the answers: $3b + 2\,a + 4\,c$
4- $z\,y$ 5-   8(9)
6- Many answers, this is only one of the answers: $2\,(8)\,(-4)$
7- Many answers, this is only one of the answers: $7\,c \,.\, 3\,y \,.\, 5\,b$
8- $y\,z\,(5 + 9 + 16) = 30\,yz$
9- $a\,b\,c\,(2 + 3 + 5) = 10\,a\,b\,c$
10- $a\,(25 - 12 - 6) = 7\,a$ 11- $x\,y\,(13 - 1 - 4) = 8\,x\,y$
12- $b\,y\,(-17 - 7 - 6) = -30\,b\,y$
13- 1   14- 0
15- a   16- 9
17- -100 18- $-x\,y\,z$
19- 99   20- $2z\,/\,b$

Exercise 2.2

1-	17 / 8		2-	139 / 70
3-	21 / 10		4-	1 / 6
5-	1		6-	1 / 6
7-	41 / 35		8-	5
9-	3		10-	157 / 120
11-	3 / 2		12-	137 / 336
13-	31 / 60		14-	99 / 100
15-	43 / 12		16-	212 / 63
17-	19 / 8		18-	37 / 18
19-	77 / 30		20-	3 / 2

Exercise 2.3

1-	1 / 3		2-	1 / 7
3-	1		4-	3 / 4
5-	2 / 3		6-	2
7-	1		8-	1
9-	3 / 5		10-	8
11-	5 / 3		12-	4 / 81
13-	5 / 4		14-	63 / 160
15-	1 / 2		16-	3 / 25
17-	7 / 18		18-	189 / 400
19-	4 / 7		20-	5 / 6

Exercise 2.4

1-	1 / 2	2-	3
3-	4 / 3	4-	4 / 15
5-	20 / 147	6-	6
7-	1	8-	34 / 9
9-	10	10-	58 / 3
11-	2 / 7	12-	2
13-	1	14-	2
15-	11 / 4	16-	1 / 2
17-	2	18-	65
19-	4	20-	10

Exercise 3.1

1-	1	2-	a^3
3-	125	4-	64
5-	$16{,}807$	6-	128
7-	$y^{1/2}$	8-	$x^{3/2}$
9-	y^8	10-	$b^{7/2}$
11-	x^2	12-	$-y^7$
13-	$-35\,b^7$	14-	$-12\,a^7\,b^4$
15-	$15\,x^7\,y^5$	16-	$(x-5)^3$
17-	$8\,x^6$	18-	$9\,x^6\,y^6$
19-	$(x-y)^3$	20-	$4\,y^5\,z^5$
21-	$(b-7)^{13}$	22-	$(2a-b)^6$
23-	$(x+y)^{20}$	24-	y^{18}
25-	$2\,y^{5/2}\,z^3$	26-	$y\,z^{1/3}$

Exercise 3.2

1- $5^{1/2}$

2- 3^3

3- $9^{7/4}$

4- 7

5- $y^{-1/4}$

6- x

7- $4x\,y$

8- $\dfrac{y^2 z^2}{3}$

9- $5\,a^2 b$

10- $2\,x\,y^{1/3}$

11- $3\,x^{1/5}\,y^{1/5}$

12- $3\,a\,b$

13- $2\,a^3 b^2 c^2$

14- $3\,y^{2/5} + y^{1/5}$

15- $\dfrac{3x^2}{2} - x^{2/3}$

16- $\dfrac{3}{x^{1/2}\,y^{2/3}}$

17- $16/9$

18- $9^{2/3} / 7^{2/3}$

19- $15^{2/3}$

20- 24

21- 3

22- $8\,a^3$

23- $\dfrac{9y^2 z^2}{16a^2 b^2}$

24- $\dfrac{10\,x^8 y^4}{5}$

25- $\dfrac{(x-y)^4}{(x+y)^4}$

26- $\dfrac{9a^2 b^2}{25y^2 z^2}$

Exercise 3.3

1- 6.4×10

2- 1.44×10^2

3- 3×10^2

4- 5.50×10^2

5- 6.25×10^2

6- 9.99×10^2

7- 1×10^3

8- 1.950×10^3

9- 7.520×10^3

10- 1.2690×10^4

11- 2.25×10^4

12- 1.679×10^5

13-	2.355×10^5	14-	5×10^5
15-	5.2×10^6	16-	7.359×10^6
17-	1.225×10^7	18-	1.256×10^8
19-	7.52×10^{-1}	20-	9.67×10^{-2}
21-	6.7×10^{-3}	22-	8.6×10^{-4}
23-	7.9×10^{-5}	24-	2.4×10^{-6}
25-	1.0×10^0	26-	2.50×10^0
27-	6.25×10^0	28-	9.749×10^0
29-	1.345×10	30-	1.795×10

Exercise 3.4

1-	0	2-	100
3-	150	4-	2,750
5-	3,240	6-	12,600
7-	56,790	8-	752,300
9-	1,190,000	10-	2,750,000
11-	74,000	12-	95,600
13-	140,000	14-	536,700
15-	454,000	16-	1,500,000
17-	6,543,200	18-	25,123,400
19-	0.61	20-	0.0985
21-	0.0022	22-	0.000678
23-	0.00003456	24-	0.00000198
25-	300,000	26-	0.00003
27-	0.04	28-	0.001666
29-	2000	30-	0.000002

Exercise 4.1

1- First degree monomial   2- First degree binomial

3- First degree trinomial, however, if like terms are added it becomes 6 y, which is a first degree monomial.

4- First degree trinomial   5- First degree multinomial

6- First degree monomial   7- Second degree binomial

8- Second degree binomial in x, first degree binomial in y

9- Second degree trinomial

10- Second degree multinomial in x, second degree multinomial in y.

11- Third degree monomial   12- Third degree binomial

13- Third degree trinomial

14- Third degree trinomial in y, first degree trinomial in x, first degree trinomial in z.

15- Fourth degree monomial   16- Fourth degree binomial

17- Fourth degree trinomial   18- Fourth degree multinomial

19- Fifth degree monomial   20- Fifth degree binomial

21- Fifth degree binomial   22- Fifth degree trinomial

23- Fifth degree multinomial   24- Sixth degree monomial

25- Sixth degree binomial   26- Sixth degree trinomial

27- Sixth degree multinomial

28- Sixth degree monomial in x, sixth degree monomial in y.

29- Sixth degree binomial in x, sixth degree binomial in y.

30- Sixth degree trinomial in a, third degree trinomial in b.

Exercise 4.2

1- $10 x + 5 y$ 2- $8 x - 2 y$

3- $5 z + 4$ 4- $7 y + 3 z$

5- $5 x + 2 y$ 6- $11 y - 5 x$

7- $5 y z + 20$ 8- $10 a b + 10 c d$

9- $4 y - 6 x$ 10- $4 x + 4 y$

11- $8a+b+1$

12- $-x-4$

13- $8x-8y+1$

14- $2y-4$

15- $4a+2b$

16- $6a+6b$

17- $3x$

18- $x+y$

19- $8ab+bc-3c$

20- $3y+3x+1$

Exercise 4.3

1- $-y$

2- $-3x$

3- $4y$

4- $-13y$

5- $6ab$

6- $12xy$

7- 0

8- $6x$

9- $6x$

10- $13x$

11- $5yz$

12- $3abc$

13- $7a+b$

14- $a-5b$

15- $5x+10y$

16- $yz+8xy$

17- $-3x-5y+9$

18- $5a+b-8$

19- $4a-11b+3c$

20- $19ab+7cd-8xy$

Exercise 4.4

1- $2x$

2- $5a^4$

3- b^5

4- $3y^8$

5- $2a^5$

6- x^4y^2

7- $-8b^5$

8- $-x^7$

9- x^3y^4

10- x^5y^5

11- $4x^8y^3$

12- $-5z^4y^2$

13- $-y^7zb^3$

14- $-54z^4y^5$

15- b^2+6b+5

16- $3x^2-10x+8$

17- $y^3-3y^2-5y+15$

18- x^3+3x^2+4x+2

19- $4x^3+18x^2+2x-24$

20- $y^4-4y^2+12y-9$

Exercise 4.5

1- $x^2 + 8x + 16$ 2- $4x^2 - 16x + 16$

3- $a^2 + 4ab + 4b^2$ 4- $4a^2 + 12ab + 9b^2$

5- $9x^2 + 6xy + y^2$ 6- $9y^2 - 6yz + z^2$

7- $4x^2 - 8xy + 4y^2$ 8- $49x^2y^2 - 28xy^2 + 4y^2z^2$

9- $16x^2 - y^2$ 10- $25b^2 - a^2$

11- $9y^2z^2 - 4x^2y^2$ 12- $-4yz$

13- $64y^4 - 96y^3 + 4y^2 + 24y + 4$

14- $x^2y^2 + 2xyz - y^2$ 15- $8abc$

16- $4a^4 - 16a^3 - 48a^2 + 128a + 256$

17- $-2b^3 - 7b^2 + 12b + 45$

18- $x^3 - 11x^2y + 25xy + 10xy^2 - 25y^3$

19- $y^4 - 8xy^3 + 24x^2y^2 - 32x^3y + 16x^4$

20- $x^4y^4 - 2x^4y^2z^2 + x^4z^4$

Exercise 4.6

1- 27 2- -4

3- 1 4- $-\frac{1}{2}$

5- 4 6- b^3

7- $-c^3$ 8- a^4

9- $(y+z)^5$ 10- $1/(a+b)^3$

11- $4y + 1$ 12- $5 - 4z$

13- $5z^2 + 2a$ 14- $-4y + 3 - (2/y)$

15- $-(3x/y) + 2 - (y/x)$ 16- $x + 2$

17- $x - 2$ 18- $x^2 + 3x - 3$

19- $x^2 - x + 2$ 20- $4x^2 + 2x - 2$

Exercise 5.1

1-	2.2.2.2.2	2-	3.3.3.3
3-	2.2.5.5	4-	2.2.2.3.5
5-	2.2.2.2.3.3	6-	3.3.5.5
7-	2.2.2.2.3.5	8-	3.7.13
9-	2.2.2.2.2.3.3	10-	2.2.2.3.3.5
11-	3.11.11	12-	2.2.3.5.7
13-	2.2.2.2.3.3.3	14-	2.2.5.5.5
15-	3.13.13	16-	2.3.3.5.7
17-	3.3.3.5.5	18-	2.7.7.7
19-	2.2.2.2.3.3.5	20-	2.2.2.3.5.7

Exercise 5.2

1-	3	2-	5
3-	10	4-	3
5-	2	6-	4
7-	21	8-	13
9-	6	10-	6
11-	6	12-	4
13-	18	14-	18
15-	7	16-	6
17-	22	18-	25
19-	21	20-	33

Exercise 5.3

1-	$a^2 b^2$	2-	$x^2 y$
3-	$y^2 z$	4-	$9(a+1$

5-	$28\,(x+3)$	6-	$z\,(z-3)$
7-	$3\,a\,(a-5)$	8-	$2\,(4\,x-3)$
9-	$5\,b^2\,(c+11)$	10-	$(y-4)$
11-	$(x-5)$	12-	$x^3\,y$
13-	$a^3\,b^2$	14-	$12\,(y-2)$
15-	$7\,(z+5)$	16-	$22\,(z+7)$
17-	$6\,(x-4)$	18-	$16\,(5\,a-3)$
19-	$25\,(5+x)^2$	20-	$(y-7)\,(5y+3)$

Exercise 5.4

1-	$5\,(y+2)$	2-	$7\,(3\,y+4)$
3-	$3\,(x-3)$	4-	$6\,(12-x)$
5-	$2\,a\,(1-3\,a)$	6-	$4\,b\,(a-2)$
7-	$a\,(3\,b+a)$	8-	$5\,a^2\,(1-2\,b)$
9-	$4\,y^2\,(3\,x+y)$	10-	$3\,x^2\,(2\,x-1)$
11-	$7x^2\,y^2\,(1-7x\,y)$	12-	$(y+2)\,(y+2+1)$
13-	$4\,(y-3)$	14-	$6(x+5)$
15-	$7\,y^3\,(1+2\,z^3)$	16-	$8\,a^2\,(1-2\,b+3\,b^2)$
17-	$9\,y^2\,z^2\,(1+3\,z-9\,z^2)$	18-	$14\,x^3\,y\,(y^2+2y-3)$
19-	$5\,x\,y^2\,z^2\,(1+2\,y\,z+3)$		
20-	$6\,x^2\,y^2\,z^2\,(y\,z^2+z\,y^2+3\,x\,z^2$		

Exercise 5.5

1-	$(y-4)\,(y+4)$	2-	$(y-7)\,(y+7)$
3-	a^2+81	4-	a^2+144
5-	$(6-z)\,(6+z)$	6-	$(8\,z-2)\,(8\,z+2)$
7-	$(4\,x-5\,y)\,(4\,x+5\,y)$	8-	$x^2\,y^2+4$
9-	$3\,(a^2-2\,a\,b)\,(a^2+2\,a\,b)$	10-	$(10\,a^2-5\,b^2)$
11-	$(7-4\,y^2)\,(7+4\,y^2)$	12-	$(6\,y^2-4\,z^2)\,(6\,y^2+4\,z^2)$

13- $(x-4)(x^2+4x+16)$ 14- $(x-6)(x^2+6x+36)$

15- $(x-1)(x^2+x+1)$ 16- $(1-3x)(1+3x+9x^2)$

17- $(2y-z)(4y^2+2yz+z^2)$

18- $2(x-3)(x^2+3x+9)$

19- $4(y+z)(y^2-2y+4)$

20- $(y+1)(y^2-y+1)$

21- $[(z+1)+y][(z+1)^2-y(z+1)+y^2]$

22- $4(3x+2y)(9x^2-6xy+4y^2)$

Exercise 5.6

1- $(x-2)(x-1)$ 2- $(x+3)(x-2)$

3- $(x+3)(x+1)$ 4- $(x-3)(x+2)$

5- $(x+4)(x+3)$ 6- $(x+4)(x-3)$

7- $(x-5)(x+2)$ 8- $(x+5)(x-3)$

9- $(x-5)(x-1)$ 10- $(x+2y)(x-y)$

11- $3(x-7)(x-2)$ 12- $3(x+y)^2$

13- $2(x-3)(x-5)$ 14- $5(x+3y)(x-y)$

15- $(2x+9)(2x-7)$ 16- $2(3x+7)(3x+5)$

17- $6(x+4)(x-3)$ 18- $3(2x-7)(2x+5)$

19- $5(2x-15)(2x-1)$ 20- $7(4x+y)(3x+2y)$

Exercise 5.7

1- $(3y-7)(3x-z)$ 2- $7(2a-b)(2x-y)$

3- $(x-3)(7x-y)$

4- $[y-(x-3z)][y+(x-3z)]$

5- $[y-(z+3)][y+(z+3)]$

6- $(x-y)(x+y)-5(z+2)$

7- $[(a+2b)-4c][(a+2b)+4c]$

8- $[(x-3y)-5z][(x-3y)+5z]$

9- $[(x+1)-2y][(x+1)+2y]$

10- $[(y-2z)-2][(y-2z)+2]$

11- $[(2a+b)-5][(2a+b)+5]$

12- $2[(a-b)+3][(a-b)-3]$

13- $[3(x+y)-4][3(x+y)+4]$

14- $5[(x-y)-3][(x-y)+3]$ 15- $(x+1)(x^2+5)$

16- $(y-1)(y^2-3)$ 17- $(2x+1)(x^2+3)$

18- $(y-3)(3y-1)(3y+1)$ 19- $(y-2)(y^2+3y+4)$

20- $(2a+b)[(2a-b)^2-3]$ 21- $(a+1)(a^2+2)$

22- $(y+1)(y^2+2)$

23- $x[(x-5)(x+5)+y(2x+y)]$

24- $(a-2)(a+2)^2$

Exercise 6.1

1-	7	2-	12
3-	$9i$	4-	3
5-	5	6-	-4
7-	x^2	8-	y^3
9-	$1/6$	10-	$5/8$
11-	0.6	12-	0.07
13-	$1/4$	14-	$3/5$
15-	0.6	16-	0.2
17-	-0.10	18-	-0.8
19-	$5^{3/4}$	20-	$y+3$
21-	$x-4$	22-	$4x^2$
23-	$z+6$	24-	$x^2 y^3$
25-	$6\sqrt{5}$	26-	$x^{5/2}$
27-	$5y^2 z^4$	28-	$z^2 y^2$

29- $y^{3/2}$

30- $a^{6/5} b^{1/5}$

31- $a^2 b^{4/3}$

32- $y^{9/4} z^{1/2}$

33- $8 a^{3/2} b^4$

34- $z^{3/2} (z - 5)$

Exercise 6.2:

1- $10\sqrt{5}$

2- $6\sqrt{6}$

3- $9\sqrt{3}$

4- $8\sqrt{7}$

5- $15\sqrt{3}$

6- $37\sqrt{5}$

7- $20\sqrt{3}$

8- $23\sqrt{2}$

9- $28\sqrt{7}$

10- $43\sqrt{3}$

11- $12\sqrt{3}$

12- $12\sqrt{5}$

13- $8\sqrt{2} + 3\sqrt{3}$

14- $8\sqrt{2} + 2\sqrt{3}$

15- $23\sqrt{2}$

16- $21\sqrt{3}$

17- 29

18- 13

19- 9

20- 19

21- $9\sqrt[3]{2}$

22- $9\sqrt[3]{3}$

23- $2\sqrt[3]{3} + 5\sqrt[3]{2}$

24- $5\sqrt[3]{5} + 2\sqrt[3]{3}$

25- $8\sqrt[3]{6} + 2\sqrt[3]{5}$

26- $3\sqrt[3]{3} + 14\sqrt[3]{2}$

27- $16\sqrt{5}$

28- $6\sqrt[3]{11}$

Exercise 6.3:

1-	$\sqrt{2}$	2-	2
3-	1	4-	$\sqrt{2}$
5-	$4\sqrt{3} - 8$	6-	1
7-	$\sqrt{6}$	8-	9
9-	0	10-	$\sqrt{7}$
11-	$-3\sqrt{3}$	12-	$\sqrt{3} - 2\sqrt{2}$
13-	-4	14-	-6
15-	$-\sqrt{3}$	16-	0
17-	5	18-	$8\sqrt[3]{2} - 2\sqrt[3]{9}$
19-	16	20-	-2
21-	4	22-	9
23-	9	24-	$-\sqrt[3]{2}$
25-	$-\sqrt[3]{3}$	26-	0
27-	$-\sqrt[3]{4}$	28-	$-\sqrt[3]{5}$

Exercise 6.4:

1-	$\sqrt{15}$	2-	$3\sqrt{6}$
3-	10	4-	7
5-	$7\sqrt{2}$	6-	$6\sqrt{2}$

7-	12	8-	$4\sqrt{3}$
9-	$9\sqrt{5}$	10-	-48
11-	$2\,y$	12-	$5\,x\,y^2\,\sqrt{x}$
13-	$8\,z\,y$	14-	$6\sqrt{2}\,x^2\,y^2$
15-	$12\,x^2\,y^2$	16-	$3\sqrt[6]{3}$
17-	x^2	18-	49
19-	$5\sqrt[12]{5^5}$	20-	$3\sqrt[3]{18}$
21-	-30	22-	40
23-	$y\sqrt[4]{y}$	24-	$y\sqrt[6]{yz^3}$
25-	$\sqrt{y^2-9}$	26-	6
27-	z^2	28-	$2\,y\,z$
29-	-37	30-	10

Exercise 6.5:

1-	2	2-	2
3-	3	4-	4
5-	$\sqrt{2}$	6-	$\sqrt{3}$
7-	$\sqrt{5}$	8-	5
9-	10	10-	2
11-	3	12-	$\sqrt[3]{3}$
13-	-3	14-	$-\sqrt[3]{15}$

15- $1/3$ 16- $x\,y$

17- a^2 18- $y\,z$

19- $2\,y$ 20- 2

21- y 22- $\dfrac{y\sqrt{2}}{2}$

23- 5 24- 6

25- $5 - \sqrt{7}$ 26- 5

27- 9 28- 5

29- $x\,y + x^2\,y^2$ <u>or</u> $x\,y\,(1 + x\,y)$

30- $x\,y - x^2\,y^2$ <u>or</u> $x\,y\,(1 - x\,y)$

31- $x\,y\,\sqrt[3]{x} + x^3\,y^2\,\sqrt[3]{y^2}$ 32- $2\,a\,b^2 + 3 + a^2\,b^2$

Exercise 6.6:

1- $\dfrac{3\sqrt{5}}{5}$ 2- $2\sqrt{2}$

3- $2\sqrt{2}$ 4- $2\sqrt{3}$

5- $\dfrac{2\sqrt[3]{3^2}}{3}$ 6- $\dfrac{5\sqrt[3]{16}}{4}$

7- $\dfrac{\sqrt[3]{36}}{6}$ 8- $2\sqrt[3]{25}$

9- $4\sqrt[4]{4}$ 10- $\dfrac{\sqrt{2y}}{y}$

11- $\dfrac{\sqrt{3z}}{z}$ 12- $\dfrac{\sqrt{y}}{y}$

13- $\dfrac{\sqrt{2xy}}{2}$ 14- $2\sqrt{2z}$

15- $\dfrac{3x\sqrt{2y}}{y}$ 16- $\dfrac{3y^{2}\sqrt{z}}{z}$

17- $\dfrac{4+\sqrt{5}}{11}$ 18- $\dfrac{3+\sqrt{3}}{3}$

19- $-\dfrac{\sqrt{5}+5}{4}$ 20- $-\sqrt{6}+3$

21- $9-3\sqrt{6}$ 22- $35+5\sqrt{42}$

23- $3\sqrt{2}+3\sqrt{3}-2-\sqrt{6}$ 24- $7-2\sqrt{10}$

25- $\dfrac{3\sqrt{2}+3\sqrt{z}+\sqrt{2z}+z}{2-z}$ 26- $\dfrac{a\sqrt{2}+\sqrt{ab}-\sqrt{2ab}-b}{2a-b}$

27- $\dfrac{\sqrt{7}+\sqrt{2}+7+\sqrt{14}}{5}$ 28- $\dfrac{z\sqrt{6}-\sqrt{6yz}-2\sqrt{yz}+2y}{2z-2y}$

Exercise 7.1

1- $13\,x$ 2- $35\,a$
3- $9\,y$ 4- $22\,z$
5- $12\,x$ 6- $30\,y$
7- $-20\,x$ 8- $-12\,a$
9- $6\,y$ 10- $-7\,z$
11- $5\,x+20$ 12- $21+9\,a$

13- $13\,y + 3$ 14- $10\,z + 2$

15- $5\,x + 3$ 16- $11\,z - 2$

17- $15\,y - 10$ 18- $19\,a - 9$

19- $-10\,x - 25$ 20- $-10\,z - 20$

Exercise 7.2

1- $x = -1$ 2- $a = 2$

3- $b = 5$ 4- $y = 3$

5- $z = 7$ 6- $x = 8$

7- $y = -12$ 8- $a = 0$

9- $y = 6$ 10- $z = 18$

11- $a = 3$ 12- $x = 5$

13- $y = 2$ 14- $z = 3$

15- $a = -3$ 16- $x = 1/2$

17- $y = 4$ 18- $b = 5$

19- $z = 2$ 20- $x = 6$

21- $a = 4$ 22- $y = 6$

23- $b = 7$ 24- $x = 5$

25- $z = -6$ 26- $y = 4$

27- $x = 2$ 28- $a = 5$

29- $y = 2$ 30- $x = 9$

31- $b = 0$ 32- $x = -1$

33- $y = -1$ 34- $z = 2$

35- $x = 4$ 36- $x = -5$

| 37- | $y = 2$ | 38- | $z = 2$ |
| 39- | $a = 11$ | 40- | $b = 2$ |

Exercise 7.3

1-	$z = 16$	2-	$y = 0.7$
3-	$x = 3$	4-	$a = 7$
5-	$y = 4$	6-	$y = -10$
7-	$x = -12$	8-	$b = 5$
9-	$z = 6$	10-	$b = 2$
11-	$a = 15$	12-	$y = 6$
13-	$x = 70$	14-	$y = 33$
15-	$z = -27$	16-	$a = 5$
17-	$b = 8$	18-	$x = -6$
19-	$y = -21$	20-	$a = 24$
21-	$z = 6$	22-	$x = 1/11$
23-	$a = 1$	24-	$y = -6/5$
25-	$z = -5$	26-	$x = 6$
27-	$b = 8$	28-	$x = -1$
29-	$y = 2$	30-	$z = -8/21$
31-	$x = 3$	32-	$a = -1$
33-	$b = 2$	34-	$y = 4$
35-	$x = -2$	36-	$y = 19/7$
37-	$y = -16$	38-	$x = 10$
39-	$z = 4$	40-	$a = -10$

Exercise 7.4

1-	25	2-	14
3-	21	4-	36, 9
5-	2, 3	6-	$1,125.00
7-	$4,800.00	8-	$2,200.00
9-	$40,000, $20,000	10-	5%, 4.5%
11-	2.5 hours	12-	4 hours
13-	74 mph, 66 mph	14-	10 mph
15-	10 miles	16-	92, 23
17-	82	18-	50, 40
19-	30	20-	24, 20, 18
21-	15 quarters, 20 dimes	22-	38 dimes, 22 nickels

23-   9 (ten dollar bills), and 7 (five dollar bills)

24-   18 (ten dollar bills), 10 (five dollar bills), and 70 (one dollar bills)

25-   24 quarters, 12 dimes, and 60 nickels

26-   length = 13 feet, width = 2 feet

27-   second side = 18 feet, third side = 6 feet

28-   Length = 14 feet, width = 7 feet

29-	6	30-	3 feet
31-	9, 11	32-	22, 24
33-	32, 34, 36	34-	25, 27, 29
35-	68, 70, 72	36-	20 gallons
37-	78%	38-	66%
39-	10%	40-	25 lbs.

Exercise 8.1

1-	$x = 0,\ x = 2$	2-	$x = 0,\ x = 5$
3-	$x = 0,\ x = 4$	4-	$x = 0,\ x = -7$
5-	$x = 0,\ x = -1$	6-	$x = 0,\ x = 3$
7-	$x = 0,\ x = 2$	8-	$x = 0,\ x = -5/2$
9-	$x = 0,\ x = -3$	10-	$x = 0,\ x = -3/2$
11-	$x = 0,\ x = -5/3$	12-	$x = 0,\ x = 9/2$
13-	$x = 0,\ x = 11/4$	14-	$x = 0,\ x = 6$
15-	$x = 0,\ x = 5/2$	16-	$x = 0,\ x = -3/2$
17-	$x = 0,\ x = -8/3$	18-	$x = 0,\ x = 9/4$
19-	$x = 0,\ x = -7/3$	20-	$x = 0,\ x = 11/4$

Exercise 8.2

1-	$x = \pm 2$	2-	$x = \pm 4$
3-	$x = \pm 5$	4-	$x = \pm 7$
5-	$x = \pm 3/2$	6-	$x = \pm \sqrt{7}$
7-	$x = \pm \sqrt{\dfrac{3}{2}}$	8-	$x = \pm \dfrac{\sqrt{3}}{2}$
9-	$x = \pm \sqrt{6}$	10-	$x = \pm 7$
11-	$x = \pm \sqrt{2}$	12-	$x = \pm \sqrt{5}$
13-	$x = \pm 3$	14-	$x = \pm \dfrac{5}{\sqrt{2}}$

15- $x = \pm \sqrt{3}$ 16- $x = \pm \dfrac{4}{\sqrt{3}}$

17- $x = \pm \sqrt{5}$ 18- $x = \pm 2\sqrt{2}$

19- $x = \pm \dfrac{4}{\sqrt{5}}$ 20- $x = \pm 2\sqrt{5}$

21- $x = \pm 2i$ 22- $x = \pm 4i$

23- $x = \pm 3i$ 24- $x = \pm 5i$

Exercise 8.3

1- $(x-3)^2 = 9$ 2- $(x-4)^2 = 16$
3- $(x-5)^2 = 25$ 4- $(x-6)^2 = 36$
5- $(x-7)^2 = 49$ 6- $(x+3)^2 = 9$
7- $(x+4)^2 = 16$ 8- $(x+5)^2 = 25$
9- $(x+6)^2 = 36$ 10- $(x+7)^2 = 49$
11- $(x+3/2)^2 = 9/4$ 12- $(x-5/2)^2 = 25/4$
13- $(x-7/2)^2 = 49/4$ 14- $(x-9/2)^2 = 81/4$
15- $(x-11/2)^2 = 121/4$ 16- $(x-3/4)^2 = 9/16$
17- $(x-1/3)^2 = 1/9$ 18- $(x-1/4)^2 = 1/16$
19- $(x-1/5)^2 = 1/25$ 20- $(x-2/7)^2 = 4/49$
21- $(x-2/3)^2 = 4/9$ 22- $(x-2/5)^2 = 4/25$
23- $(y-5)^2 = 9$ 24- $(x+2)^2 = 16$
25- $(z-3)^2 = 8$ 26- $(y-5/2)^2 = 1/4$
27- $(a+½)^2 = 1$ 28- $(z-1)^2 = 9$
29- $(b+1/6)^2 = 37/36$ 30- $(y-1/3)^2 = 4/9$
31- $(y+1)^2 = 5$ 32- $(a+2)^2 = 13$
33- $(z+3)^2 = 25$ 34- $(b+4)^2 = 49$
35- $(y+5)^2 = 64$ 36- $(x-1)^2 = 5$
37- $(a-2)^2 = 8$ 38- $(z-3)^2 = 25$
39- $(b-4)^2 = 49$ 40- $(x-5)^2 = 64$

Exercise 8.4

1-	$x = \pm 4$	2-	$x = \pm 5$
3-	$x = \pm 6$	4-	$x = \pm 7$
5-	$x = \pm 8$	6-	$x = 0,\ x = 1$
7-	$x = 0,\ x = 2$	8-	$x = 0,\ x = 3$
9-	$x = 0,\ x = 4$	10-	$x = 0,\ x = 5$
11-	$y = 0,\ y = 2$	12-	$y = 0,\ y = 3$
13-	$y = 0,\ y = 4$	14-	$y = 0,\ y = 5$
15-	$y = 0,\ y = 6$	16-	$y = 0,\ y = 2/3$
17-	$y = 0,\ y = 4/3$	18-	$y = 0,\ y = 1/4$
19-	$y = 0,\ y = 1/2$	20-	$y = 0,\ y = 5/4$
21-	$x = 1,\ x = 2$	22-	$x = 1,\ x = 4$
23-	$x = 3,\ x = 4$	24-	$x = 3 - \sqrt{3},\ x = 3 + \sqrt{3}$
25-	$x = 1,\ x = 5$	26-	$x = 2,\ x = 5$
27-	$x = 4,\ x = 5$	28-	$x = 3,\ x = 5$
29-	$x = 1$	30-	$x = 3$
31-	$y = 4,\ y = 6$	32-	$y = 3/2,\ y = 6$
33-	$y = 2,\ y = 6$	34-	$y = 3,\ y = 6$
35-	$y = 5,\ y = 6$	36-	$y = 5$
37-	$y = 6$	38-	$y = 1,\ y = 3$
39-	$y = 5,\ y = -6$	40-	$y = 2,\ y = 3$
41-	$x = -1/2,\ x = -2$	42-	$x = -1,\ x = -3$
43-	$x = 1,\ x = -4$	44-	$x = 2,\ x = -5$
45-	$x = 4/3,\ x = -6$	46-	$x = -2,\ x = -5/2$
47-	$x = 1,\ x = -2$	48-	$x = 1,\ x = -3/2$
49-	$x = -2,\ x = -5/2$	50-	$x = 3/2,\ x = -3$

Exercise 9.1

1-	$x < 3$	2-	$x > 3$
3-	$x < 4$	4-	$x > 5$
5-	$x \leq 11$	6-	$x \geq 1$
7-	$x \geq -1$	8-	$x \leq -10$

9-	$x < -3$		10-	$x > -2/3$
11-	$x > 3$		12-	$x < 2$
13-	$x > 3/4$		14-	$x < 4$
15-	$x \geq 1$		16-	$x \leq -9$
17-	$x < -1$		18-	$x \geq -1$
19-	$x < 1$		20-	$x > -4/3$
21-	$x \leq -3$		22-	$x > -4$
23-	$x \geq 16/9$		24-	$x \geq -8/7$
25-	$x > 4$		26-	$x < 6$
27-	$x > 2$		28-	$x < 2$
29-	$x \geq -1$		30-	$x \leq 4$
31-	$x \geq 2$		32-	$x \leq 1/2$
33-	$x \geq -1$		34-	$x \leq -1$
35-	$x > 1$		36-	$x < 2$
37-	$x < 17/5$		38-	$x > -12/5$
39-	$x \leq 13/3$		40-	$x \leq -5/4$

Exercise 9.2

1-	$x > 6, x < 2$		2-	$x < 3, x > -1$
3-	$x > 9, x < -5$		4-	$x < -2, x > -8$
5-	$x \geq 3, x \leq -19$		6-	$x \leq 11, x \geq 1$
7-	$x > 9, x < -3$		8-	$x < 8, x > 6$
9-	$x > 4, x < -4$		10-	$x < -8, x > -4$
11-	$x \geq 5, x \leq -1$		12-	$x \leq 3/2, x \geq -15/2$
13-	$x \geq -2, x \leq 10/3$		14-	$x \leq -4, x \geq 14/3$
15-	$x \geq 7/2, x \leq -5$		16-	$x \leq -1, x \geq 9/5$

17- $x > 1/2,\ x < -1/6$ 18- $x < -3/4,\ x > 13/4$

19- $x > -10/7,\ x < 8/7$ 20- $x < -13/3,\ x > 3$

21- $x \geq 4/5,\ x \leq 2/5$ 22- $x \leq -5/8,\ x \geq -1/8$

23- $x > 1/10,\ x < -9/10$ 24- $x < 8/9,\ x > 0$

25- $x < -1/3,\ x < -5$ 26- $x < 3,\ x > 3$

27- $x > 6,\ x < -4$ 28- $x < -9,\ x > 1/3$

29- $x \geq 10$ 30- $x \leq 1/5,\ x \leq 5/2$

31- $x > 1/5,\ x > 3/5$ 32- $x < 5/6,\ x > -11/4$

33- $x > 2,\ x < -4$ 34- $x < 1/3,\ x > -7/3$

35- $x \geq 1,\ x \geq -3$ 36- $x < 10/7,\ x > -4$

37- $x \leq 1/5,\ x \geq 0$ 38- $x > 4/5,\ x < 1/3$

39- $x < 5/12,\ x > 3/10$ 40- $x > -2/9,\ x > -16/5$

Exercise 10.1

Problems No. 1, 2, 3, 4 & 5

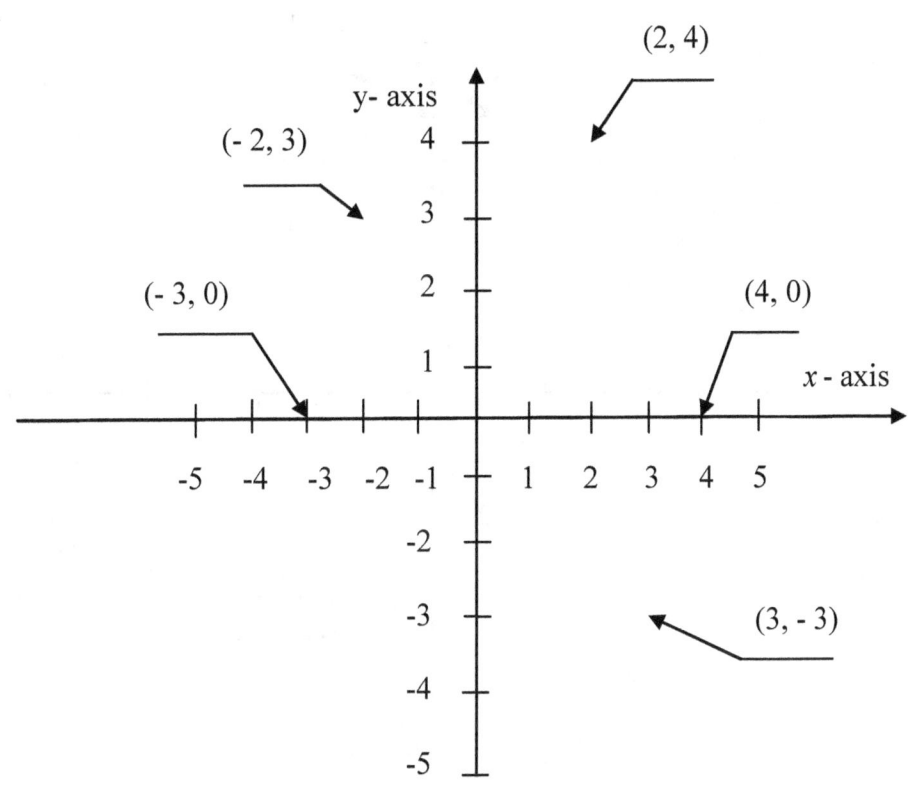

Exercise 10.1

Problems No. 6, 7, 8, 9 & 10

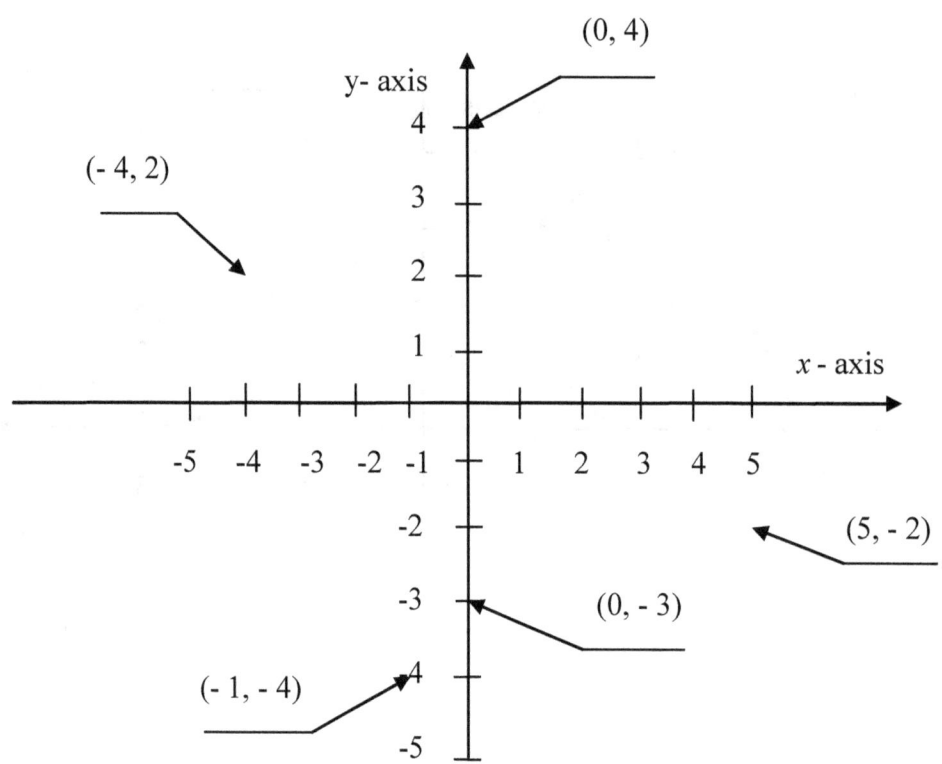

Exercise 10.1

Problems No. 11, 12, 13 & 14

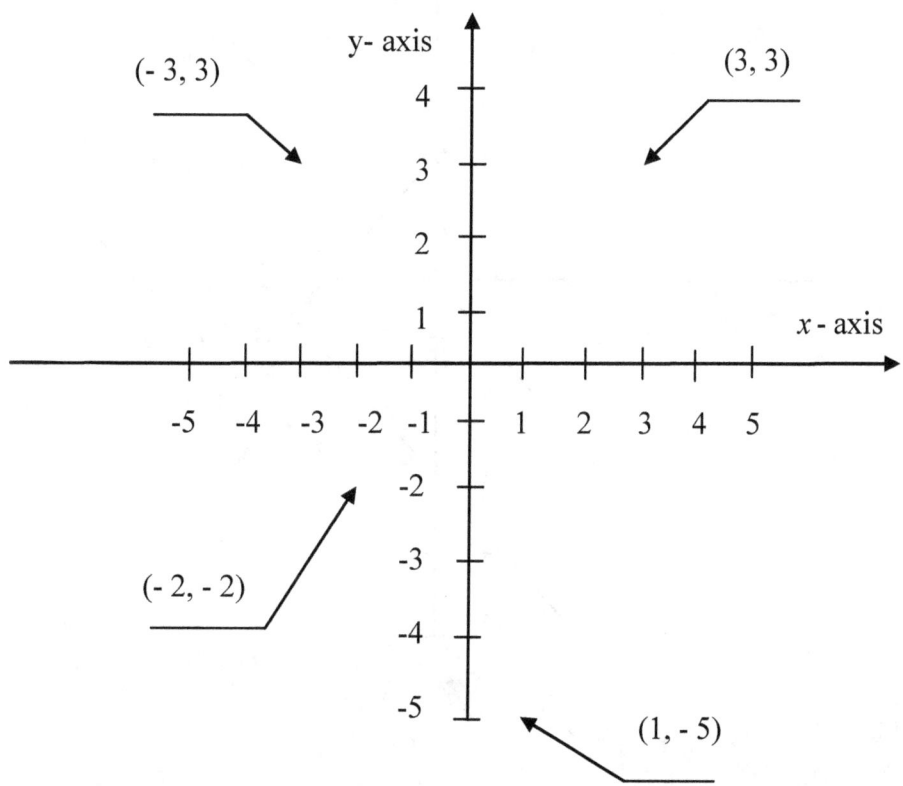

Exercise 10.2

Problems No. 1, 2, & 3

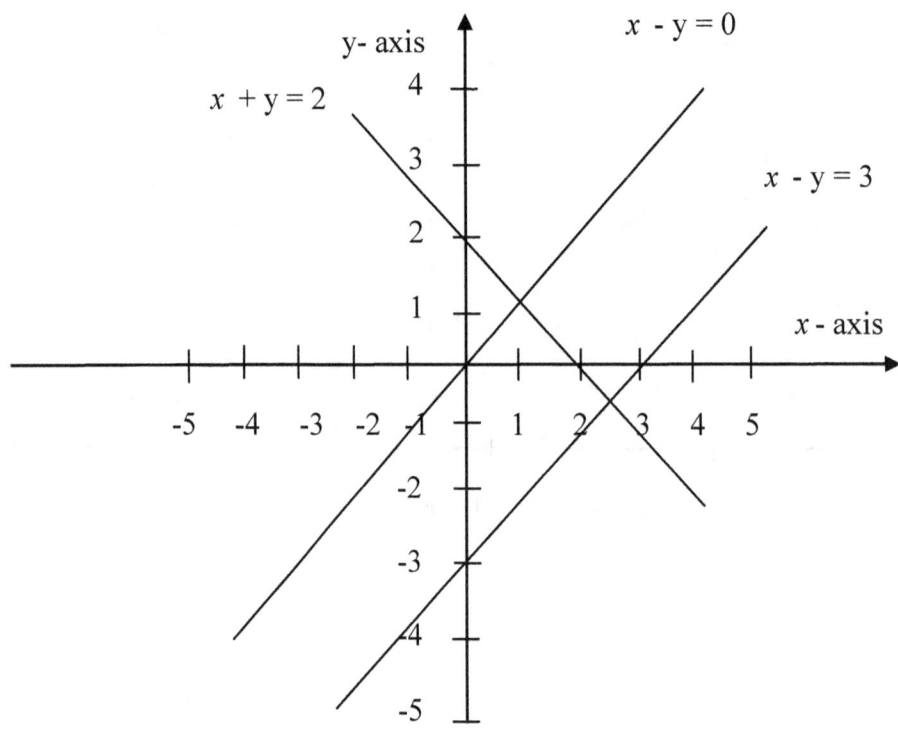

Exercise 10.2

Problems No. 4, 5, & 6

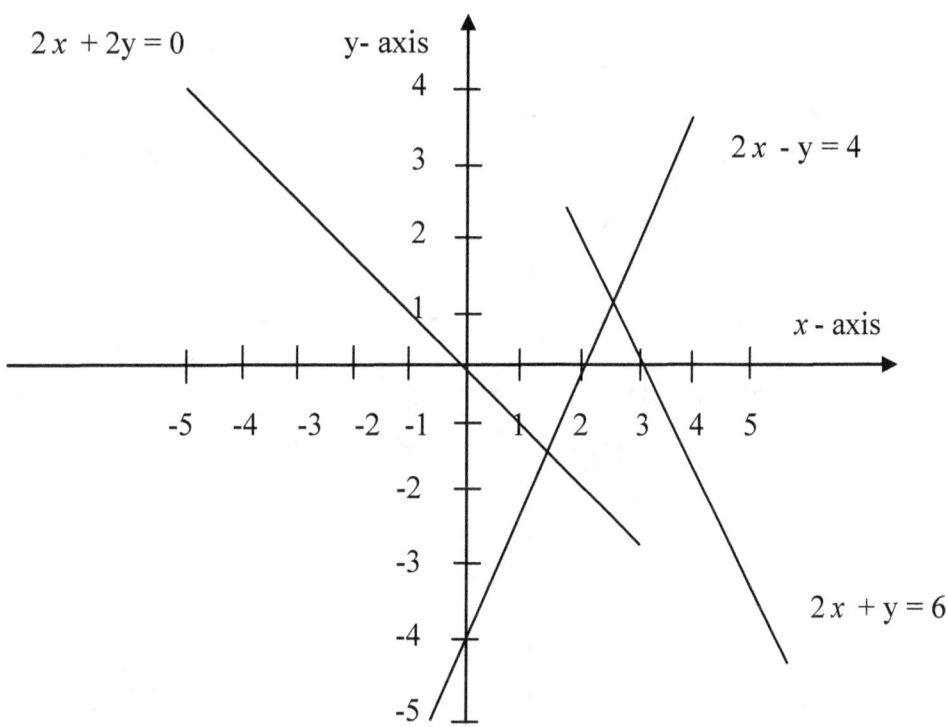

Exercise 10.2

Problems No. 7, 8, 9 & 10

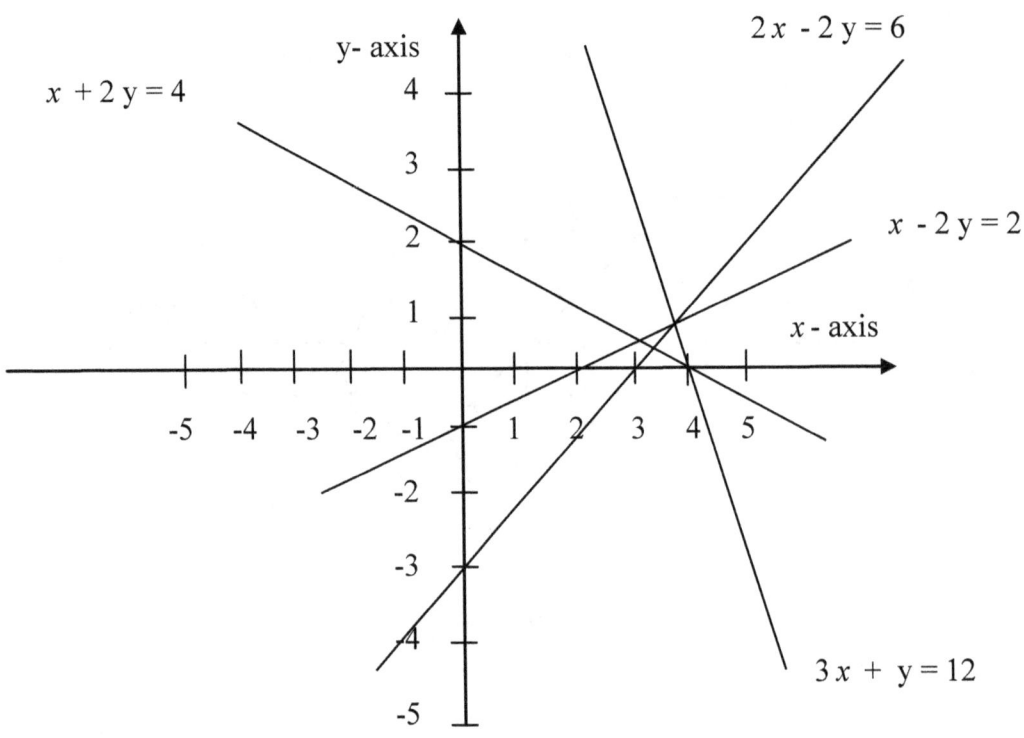

Exercise 10.2

Problems No. 11 & 12

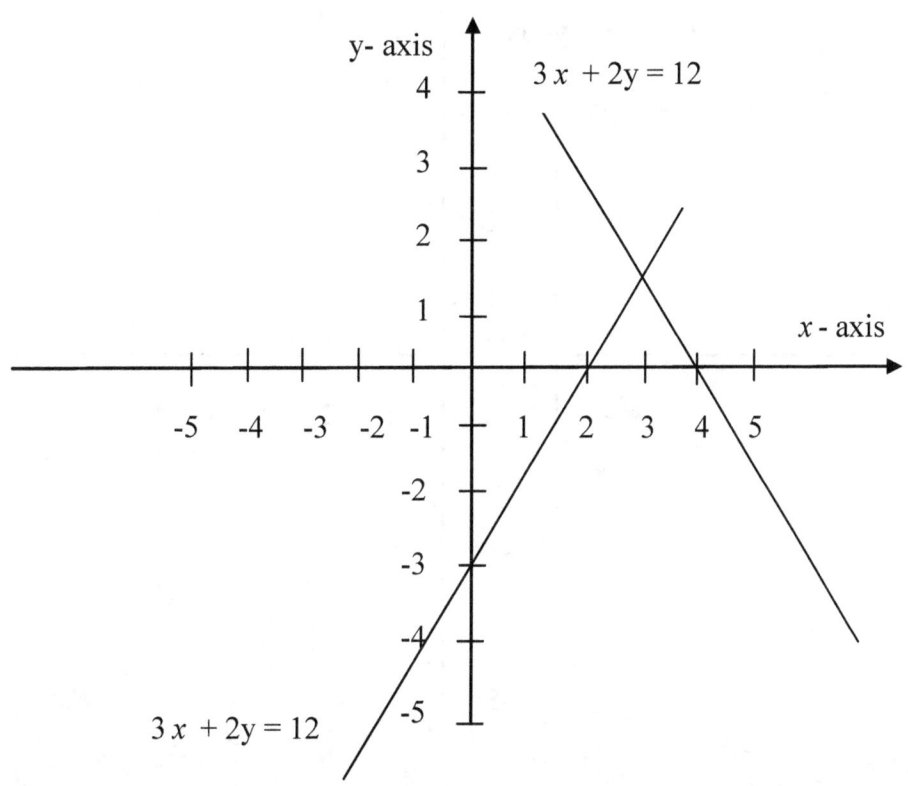

Exercise 10.2

Problems No. 13 & 14

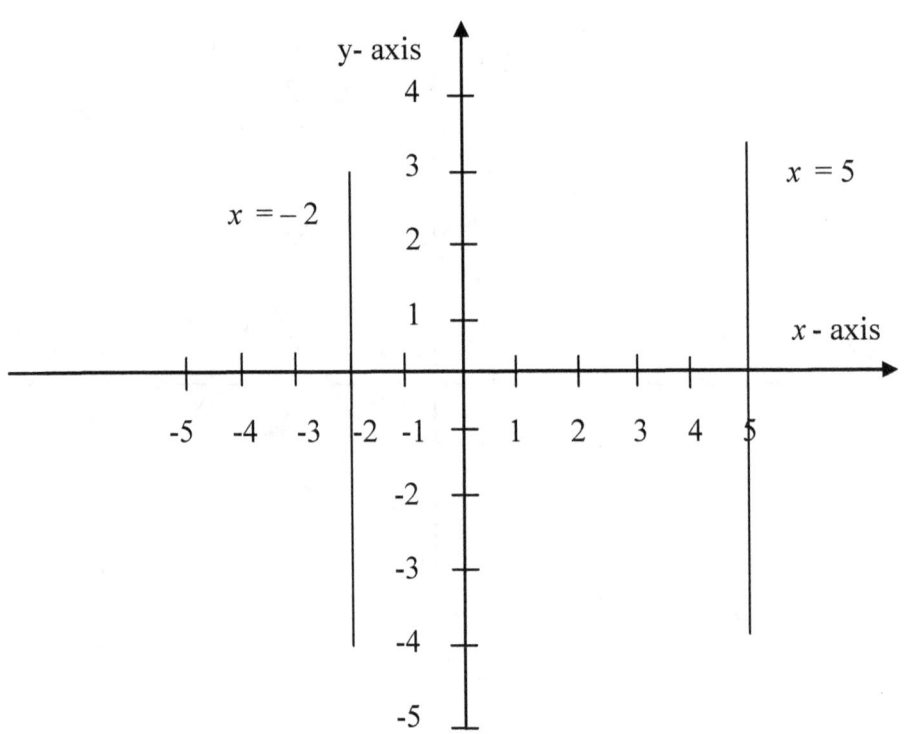

Exercise 10.2

Problems No. 15 & 16

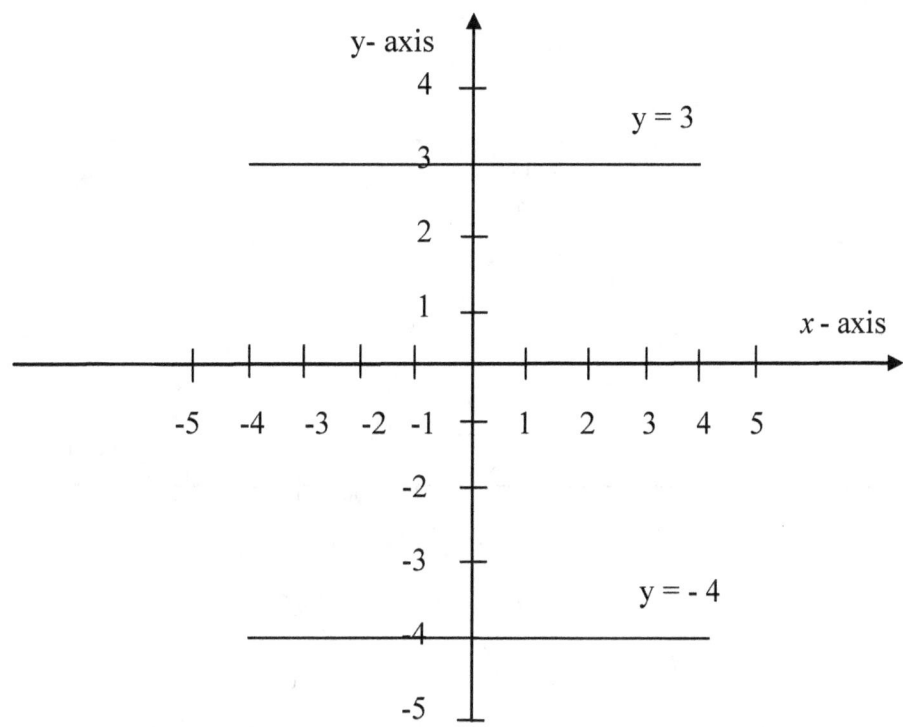

Exercise 10.2

Problems No. 17, 18, 19 & 20

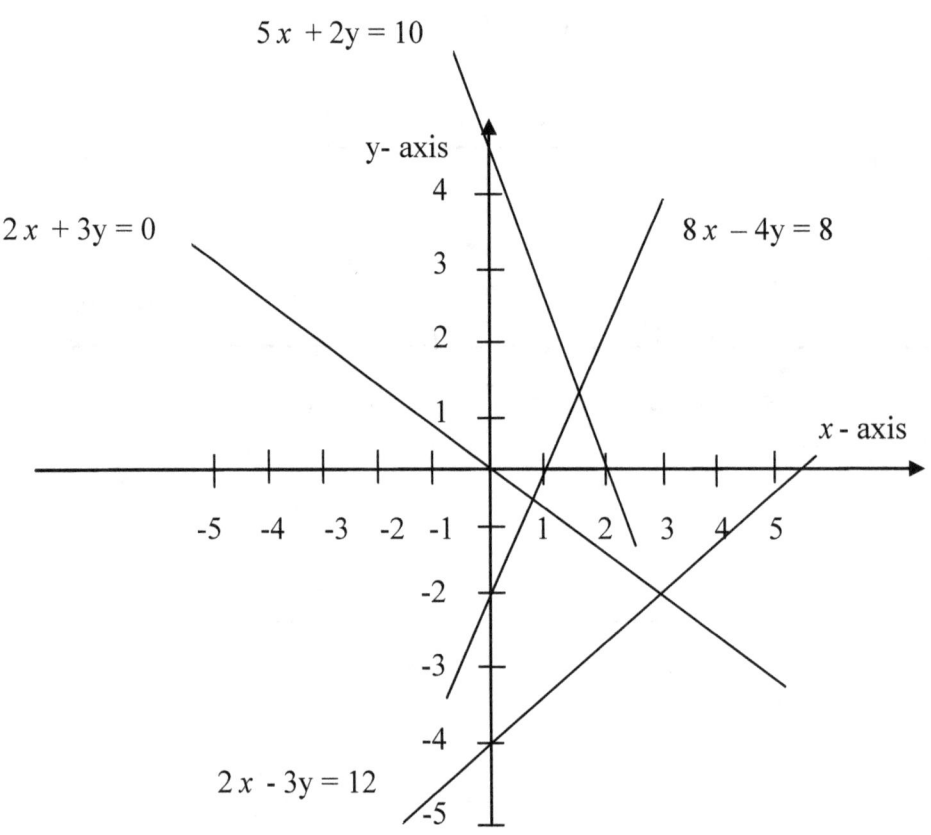

Exercise 10.2

Problems No. 21, 22, 23 & 24

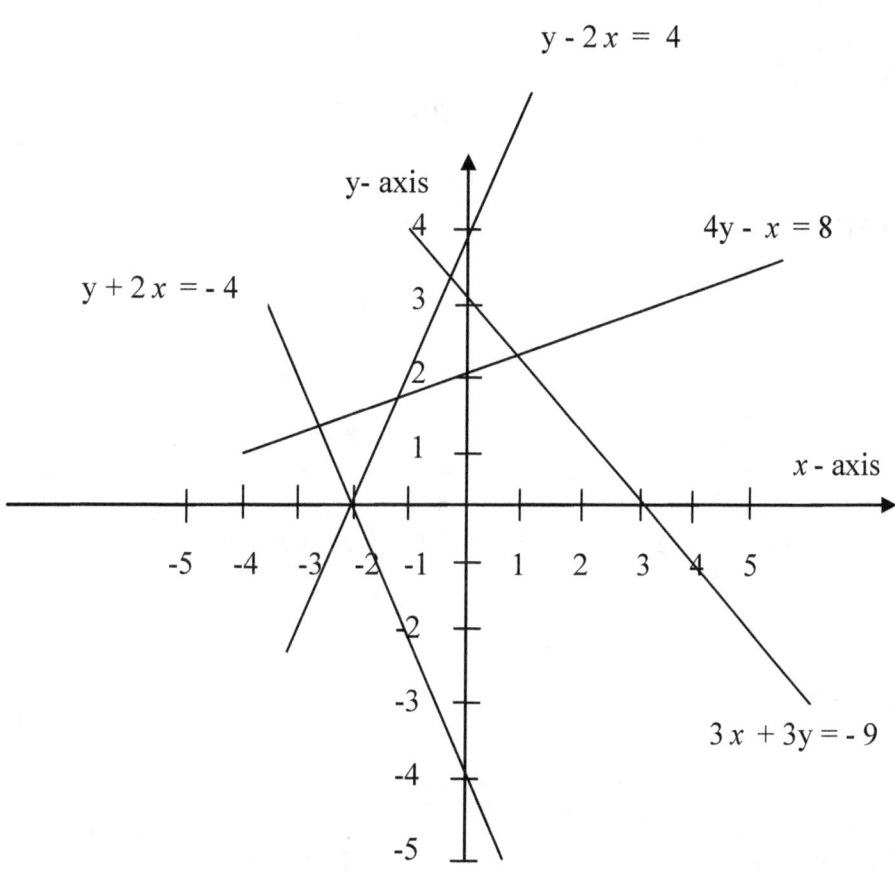

Exercise 10.3

1-	1	2-	-1
3-	1/2	4-	1/2
5-	$-1/2$	6-	1/3
7-	2	8-	-1
9-	2/3	10-	2
11-	3/2	12-	1
13-	1	14-	1/3
15-	1	16-	4/3
17-	$-1/3$	18-	-1
19-	$-1/4$	20-	1/5

Exercise 10.4

A-

1-	$y = x - 1$	2-	$y = 3x - 7$
3-	$y = -2x + 2$	4-	$y = -3x + 9$
5-	$y = -4x - 11$	6-	$y = 2x - 10$
7-	$y = 3x/2 + 17/2$	8-	$y = x - 6$
9-	$y = -3x - 13$	10-	$y = 4x + 4$

Exercise 10.4

B-

11-	$y = x$	12-	$4y = 3x + 4$
13-	$2y = 3x + 1$	14-	$3y = x + 7$
15-	$y = 4$	16-	$y = x + 2$
17-	$7y = 5x + 6$	18-	$4y = 5x + 1$
19-	$y = -2x - 1$	20-	$y = -4x$

Exercise 11.1

1-	$x = 3, y = 3$	2-	$x = 4, y = -2$
3-	$x = 6, y = 2$	4-	$x = 2, y = 3$
5-	$x = 3, y = 5$	6-	$x = 1, y = 4$
7-	$x = 0, y = 5$	8-	$x = 2, y = 4$
9-	$x = 4, y = 0$	10-	$x = -1, y = 3$
11-	$x = -1, y = -1$	12-	$x = -1, y = 2$
13-	$x = 0, y = -3$	14-	$x = 1, y = 5$
15-	$x = 3, y = 1$	16-	$x = 4, y = 2$
17-	$x = 5, y = 0$	18-	$x = 5, y = 1$
19-	$x = 4, y = 1$	20-	$x = 3, y = 0$
21-	$x = 4, y = -1$	22-	$x = 4, y = 3$
23-	$x = 1, y = 3$	24-	$x = 0, y = 6$

Exercise 11.2

1-	$x = 6, y = 3$	2-	$x = 6, y = 2$
3-	$x = 8, y = 2$	4-	$x = 1, y = 2$
5-	$x = 0, y = 4$	6-	$x = 3, y = 4$
7-	$x = 3, y = 2$	8-	$x = 2, y = 2$
9-	$x = 4, y = 4$	10-	$x = 5, y = 2$
11-	$x = 3, y = 5$	12-	$x = 1, y = 5$
13-	$x = 0, y = 4$	14-	$x = 4, y = 1$
15-	$x = 6, y = 1$	16-	$x = 2, y = 5$
17-	$x = 1, y = 3$	18-	$x = 2, y = 4$
19-	$x = 5, y = 1$	20-	$x = 3, y = 1$
21-	$x = 3, y = 3$	22-	$x = 4, y = 5$
23-	$x = 6, y = 2$	24-	$x = 1, y = 6$

Exercise 11.3

Problem No. 1

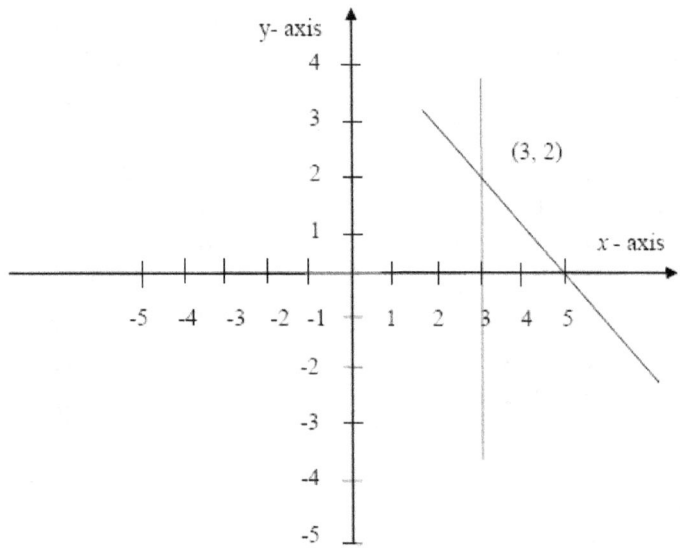

Problem No. 2

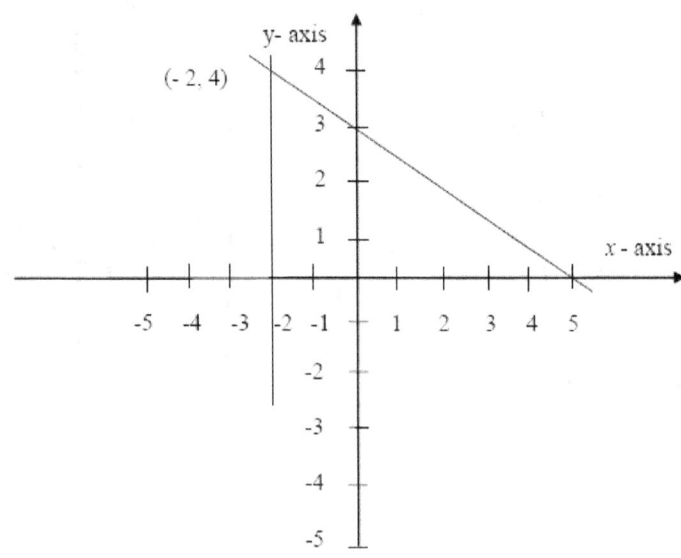

Exercise 11.3

Problem No. 3

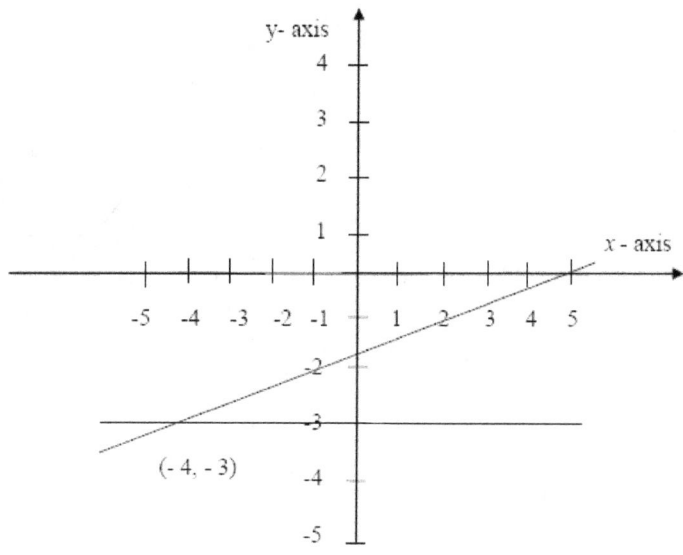

Problem No. 4

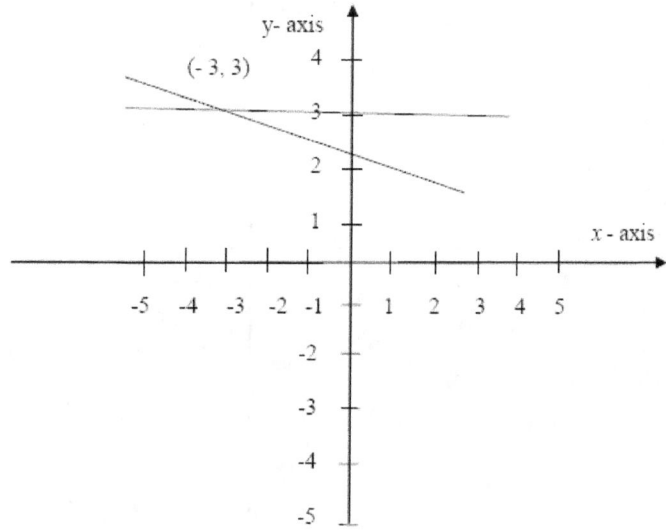

Exercise 11.3

Problem No. 5

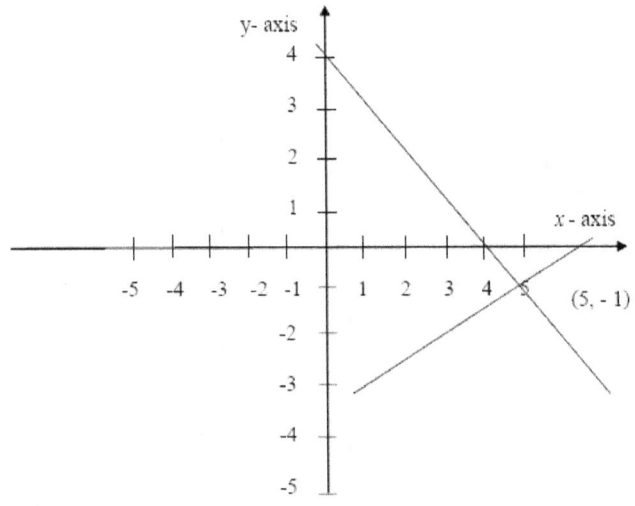

Problem No. 6

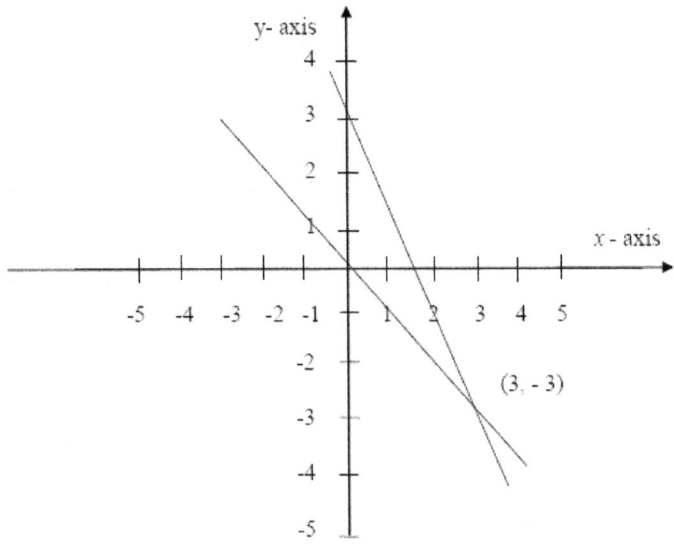

Exercise 11.3

Problem No. 7

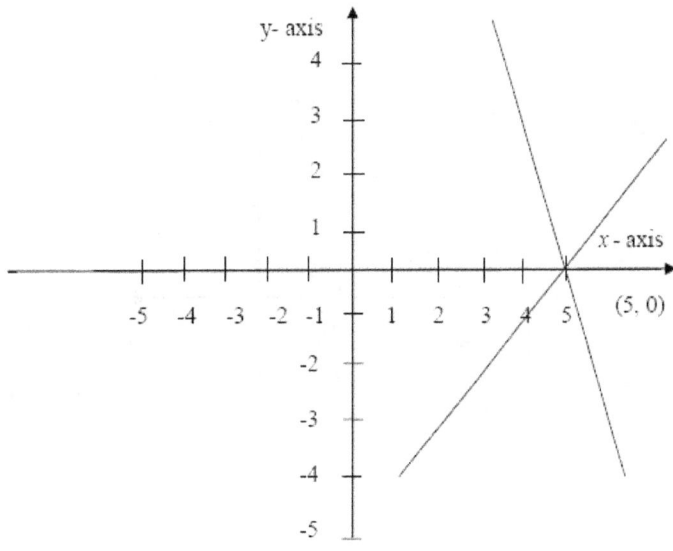

Problem No. 8

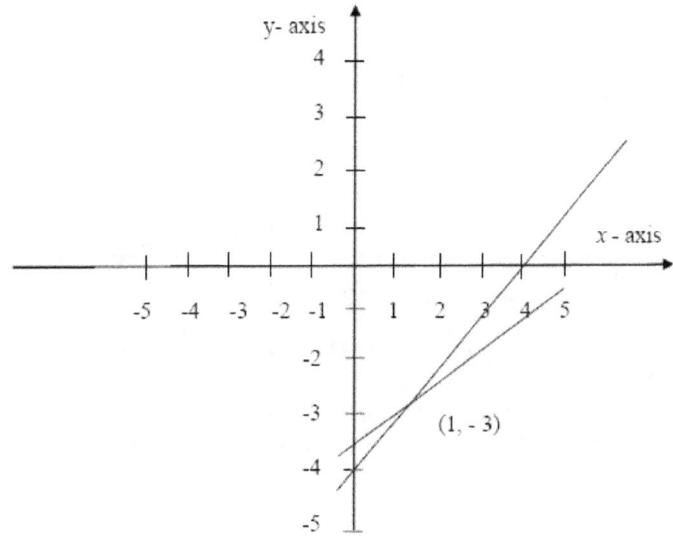

Exercise 11.3

Problem No. 9

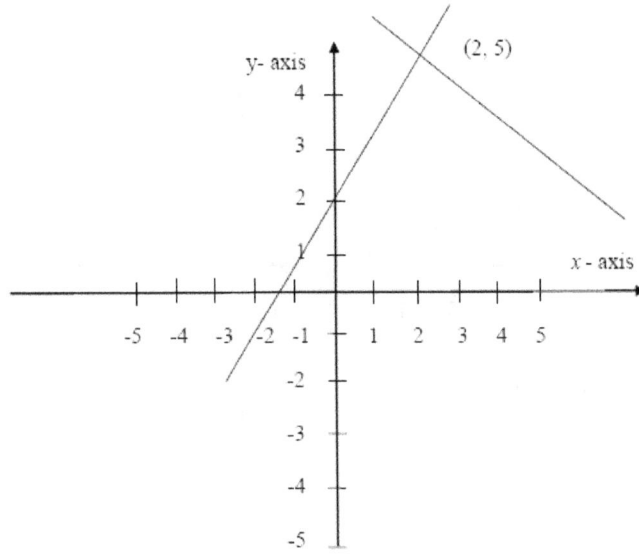

Problem No. 10

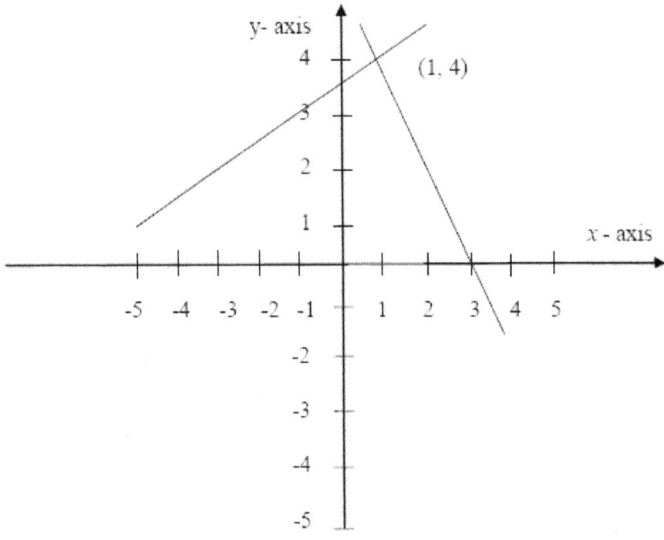

Exercise 11.3

Problem No. 11

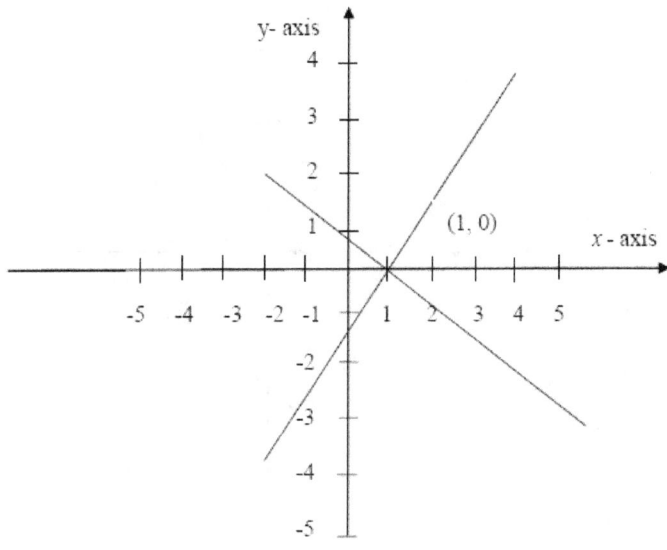

Problem No. 12

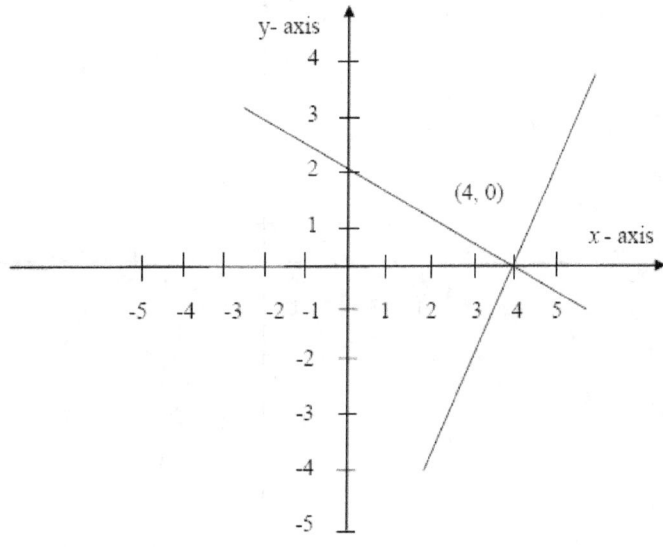

Exercise 11.3

Problem No. 13

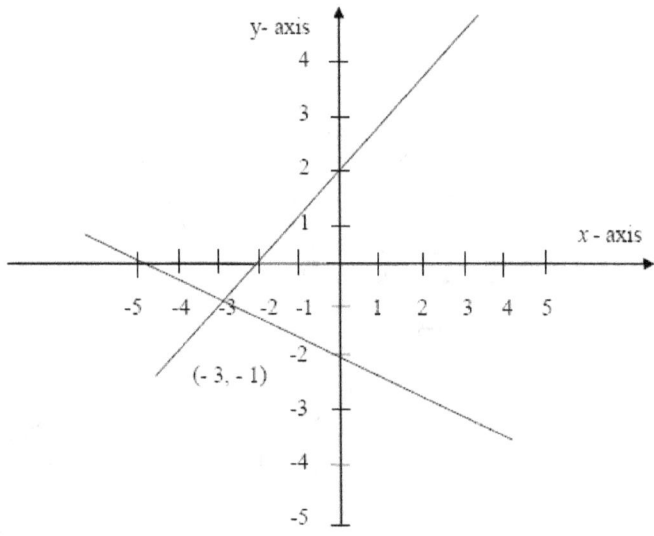

Problem No. 14

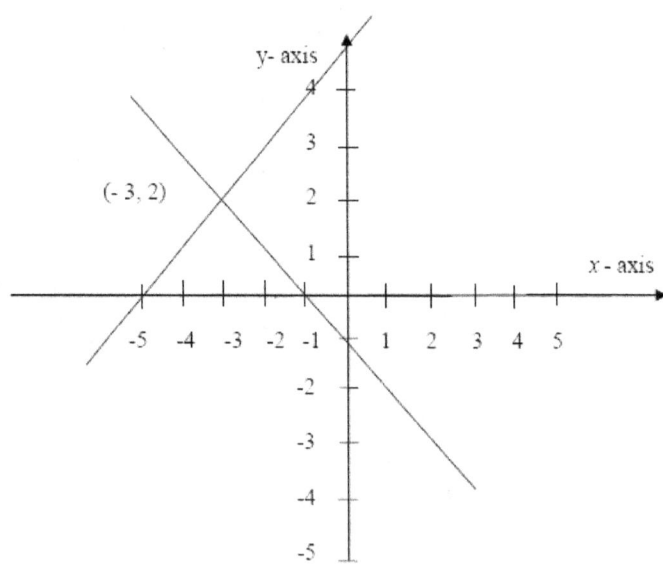

Exercise 11.3

Problem No. 15

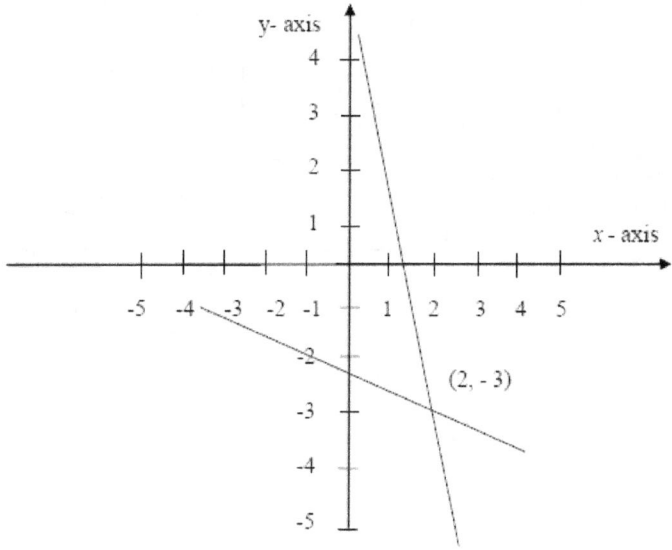

Problem No. 16

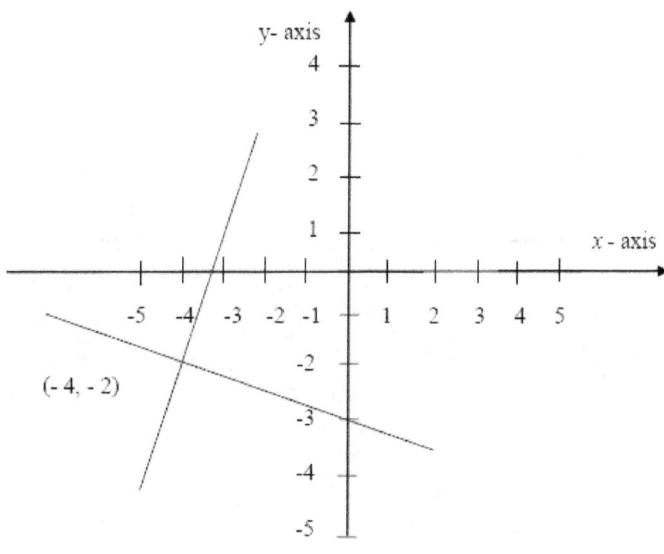

Exercise 11.3

Problem No. 17

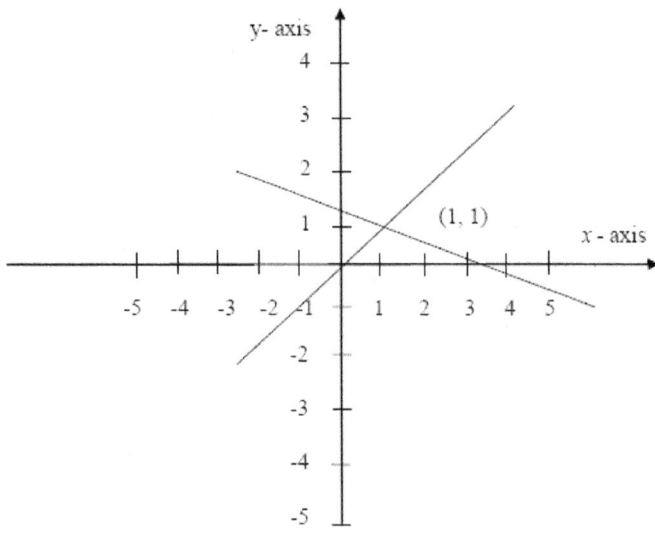

Problem No. 18

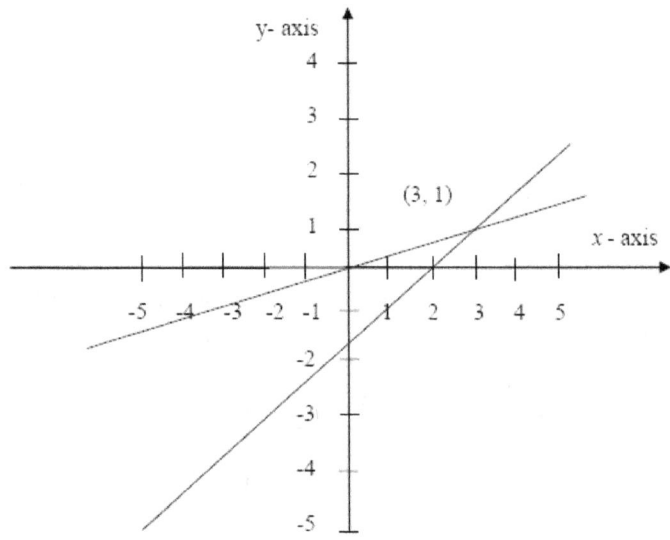

Exercise 11.3

Problem No. 19

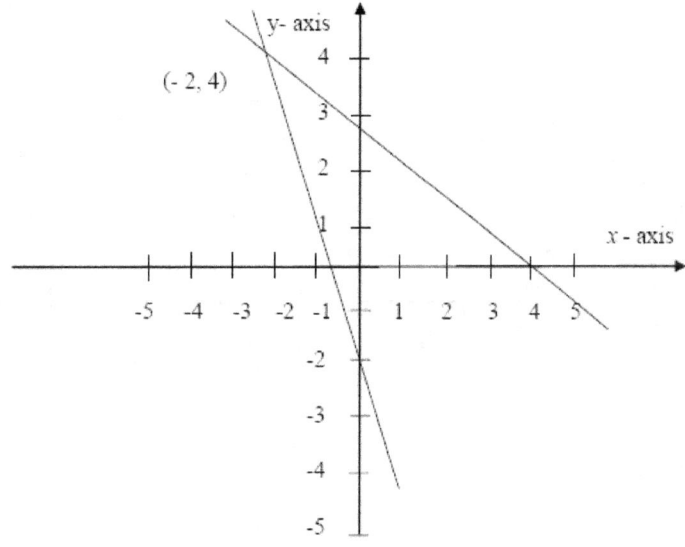

Problem No. 20

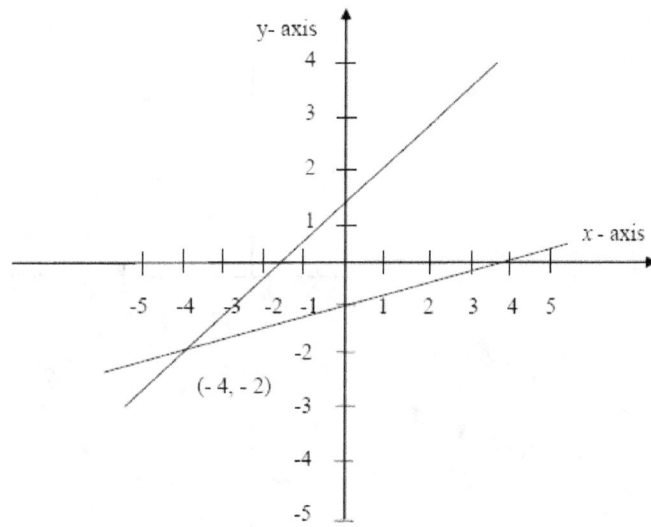

Exercise 11.3

Problem No. 21

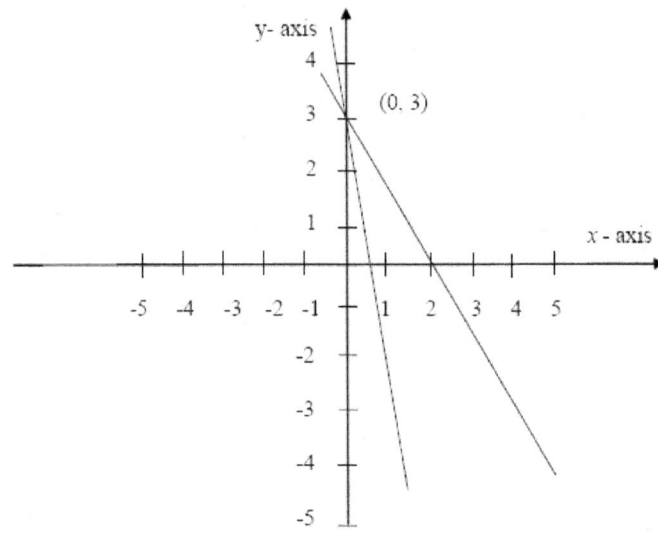

Problem No. 22

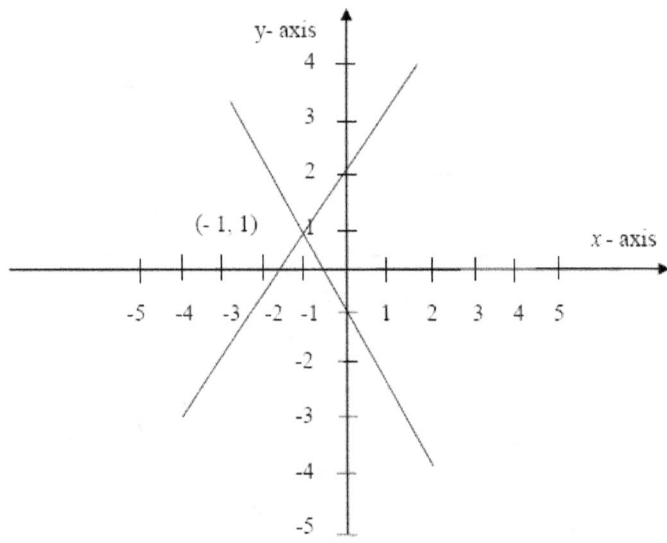

Exercise 11.3

Problem No. 23

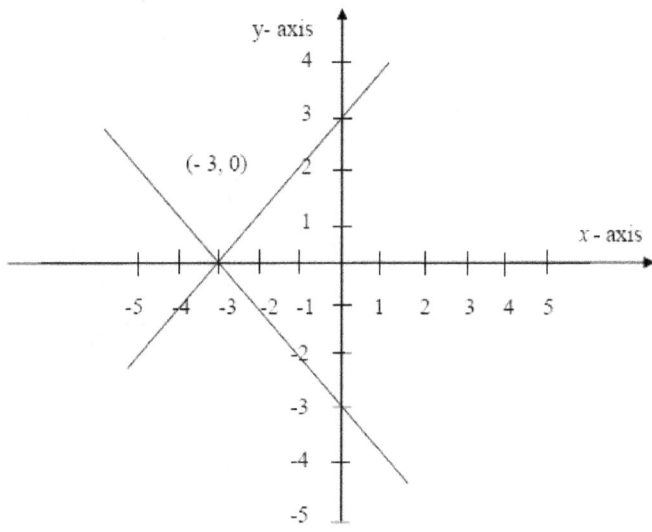

Problem No. 24

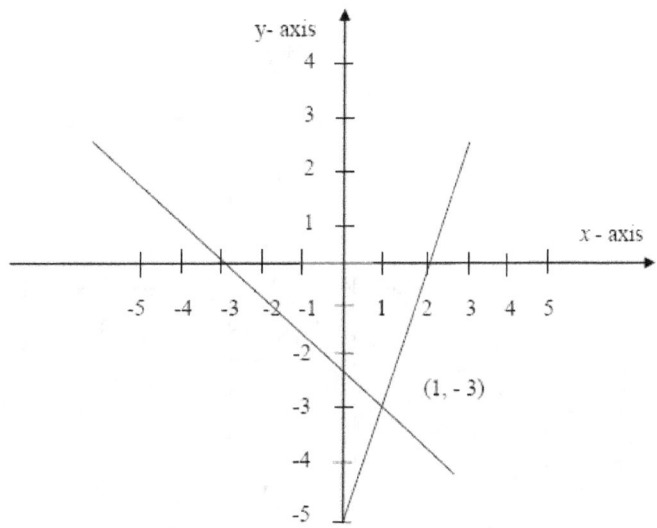

Exercise 11.4

Problem No. 1

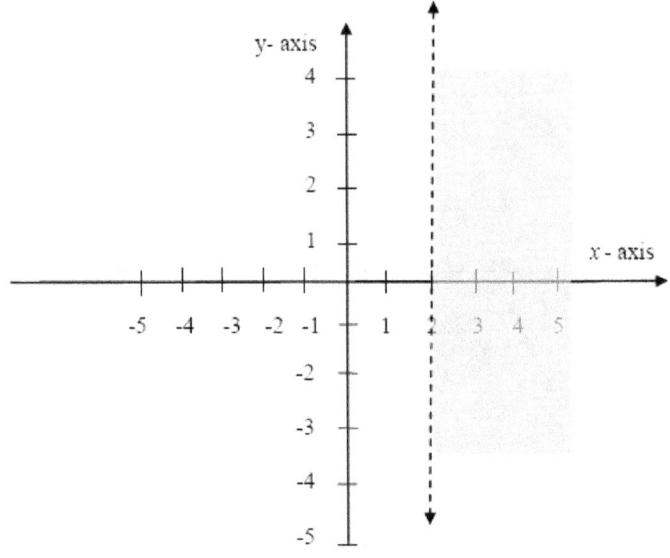

Problem No. 2

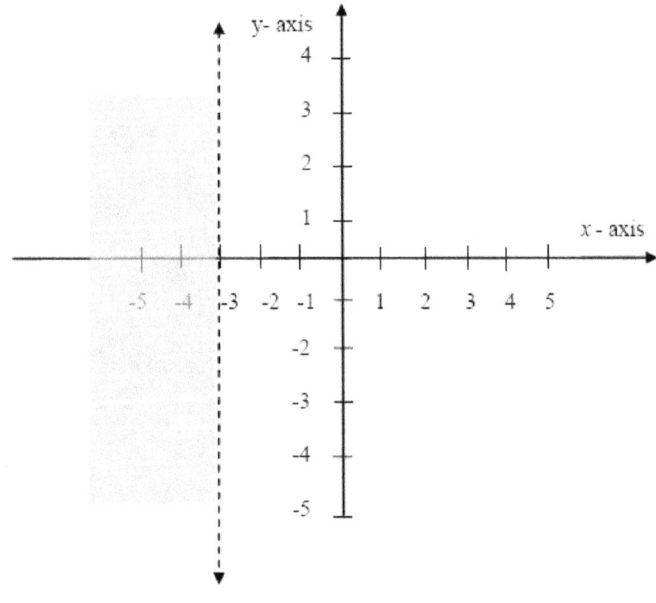

Exercise 11.4

Problem No. 3

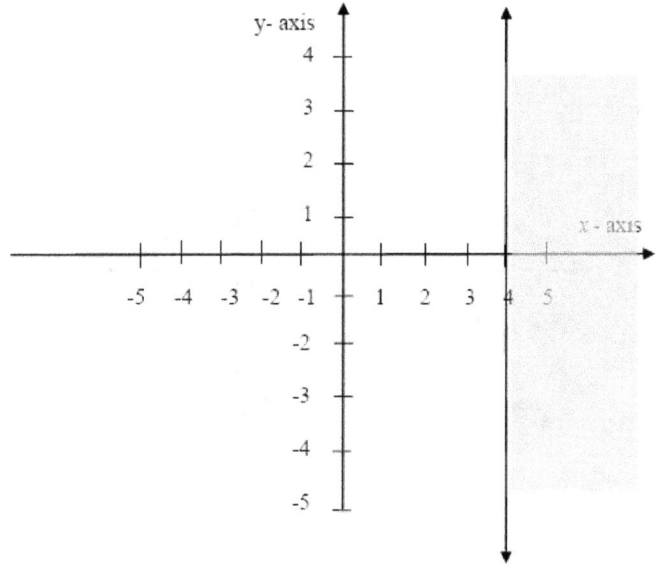

Problem No. 4

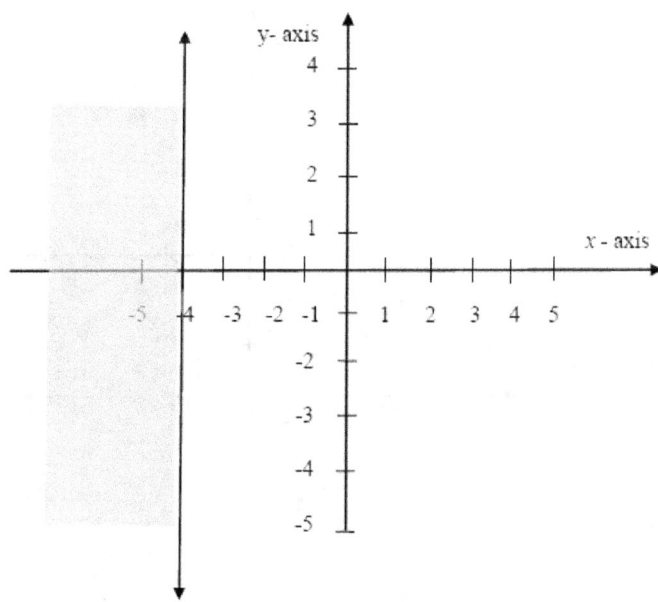

Exercise 11.4

Problem No. 5

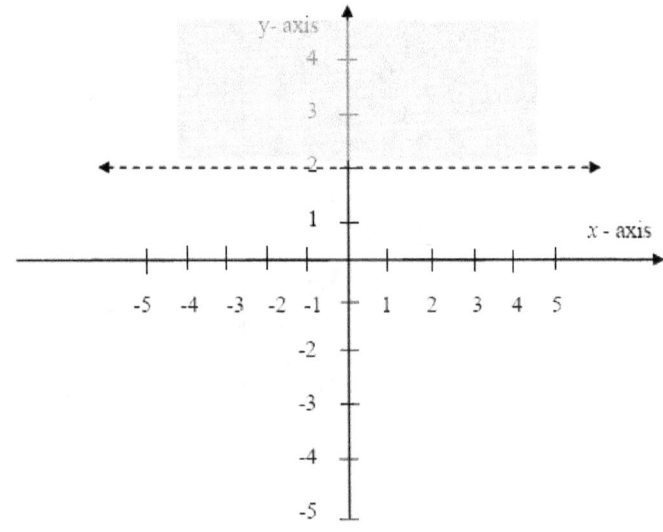

Problem No. 6

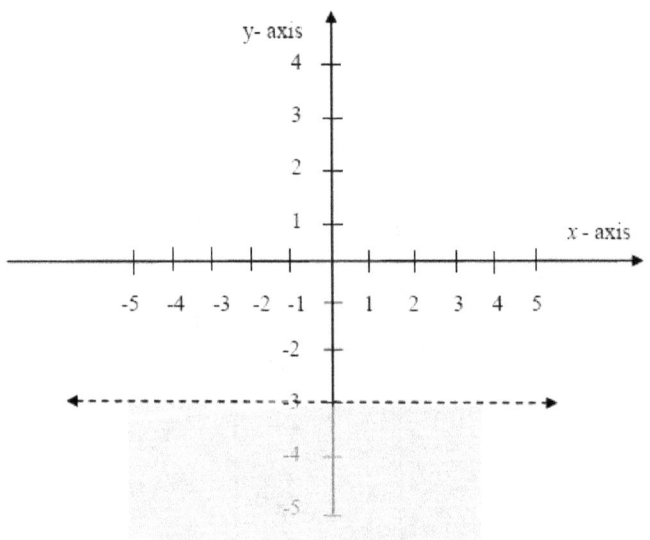

Exercise 11.4

Problem No. 7

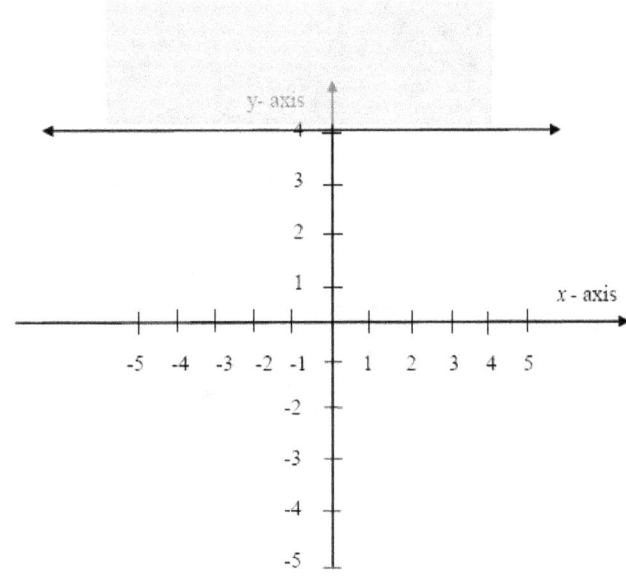

Problem No. 8

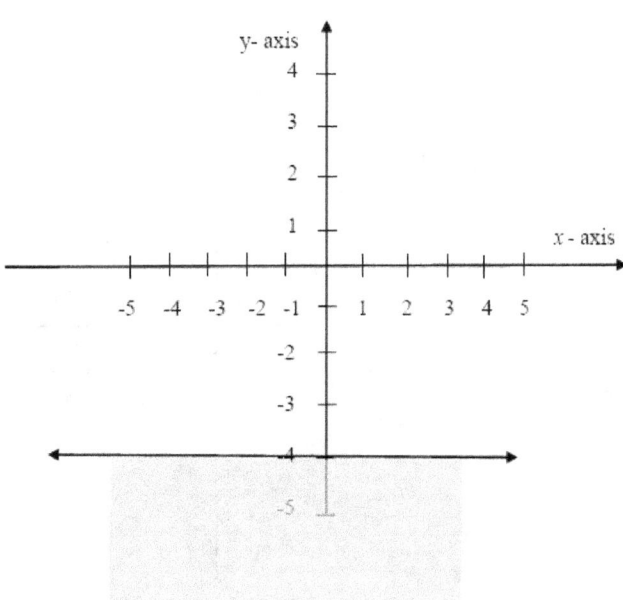

Exercise 11.4

Problem No. 9

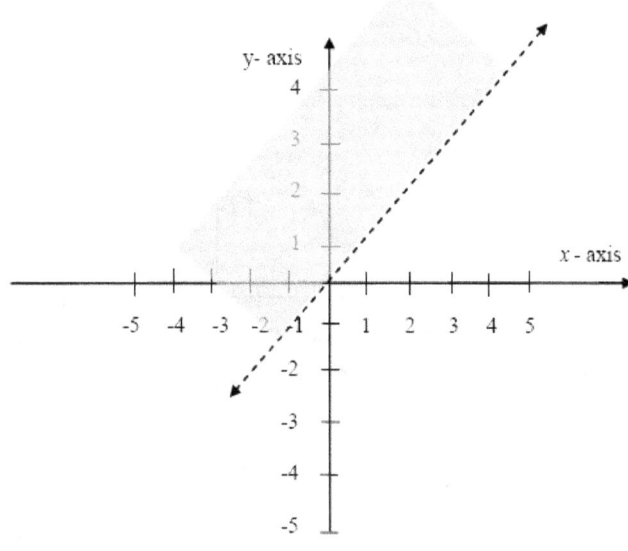

Problem No. 10

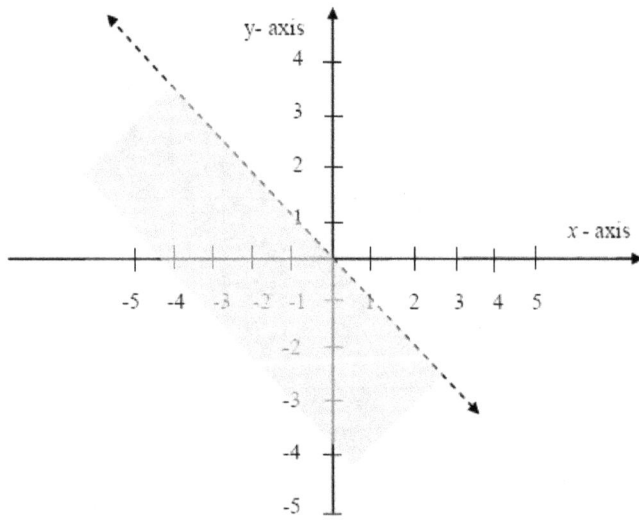

Exercise 11.4

Problem No. 11

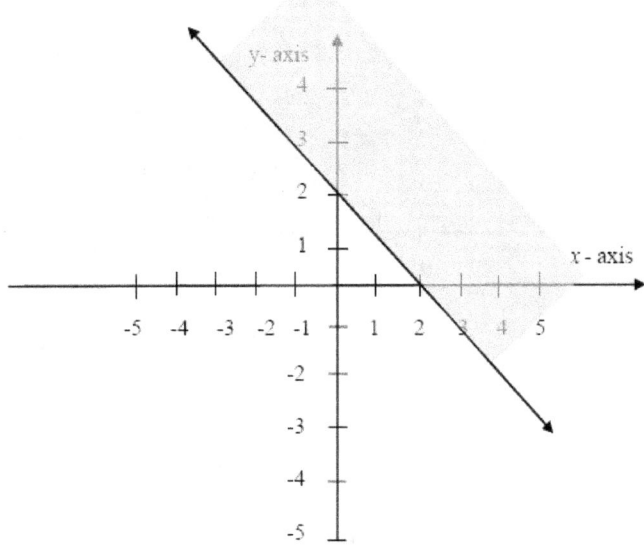

Problem No. 12

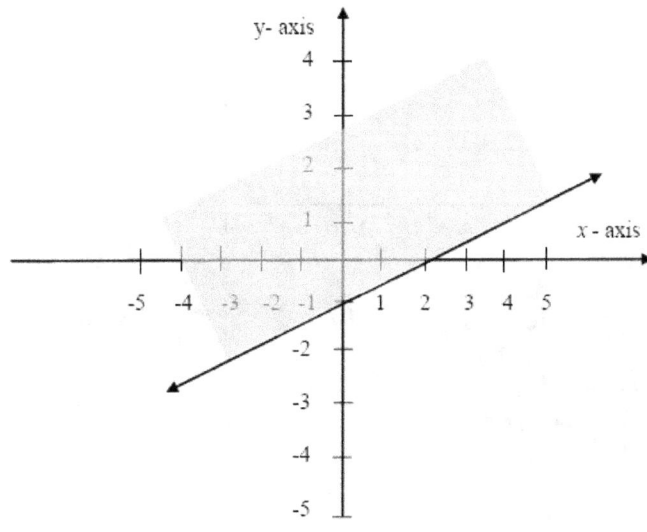

Exercise 11.4

Problem No. 13

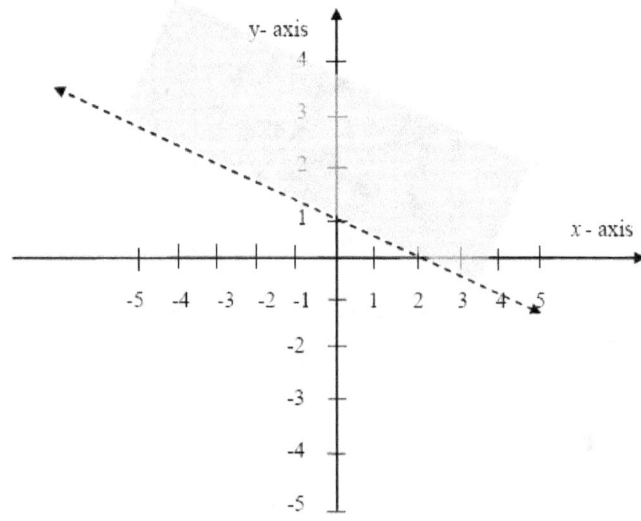

Problem No. 14

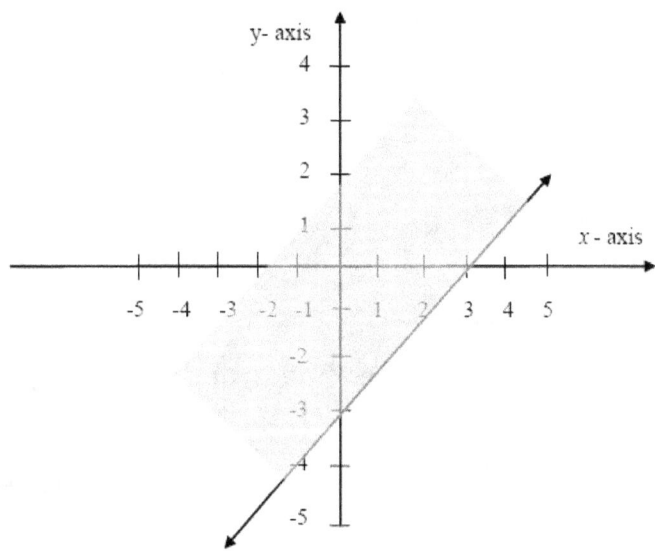

Exercise 11.4

Problem No. 15

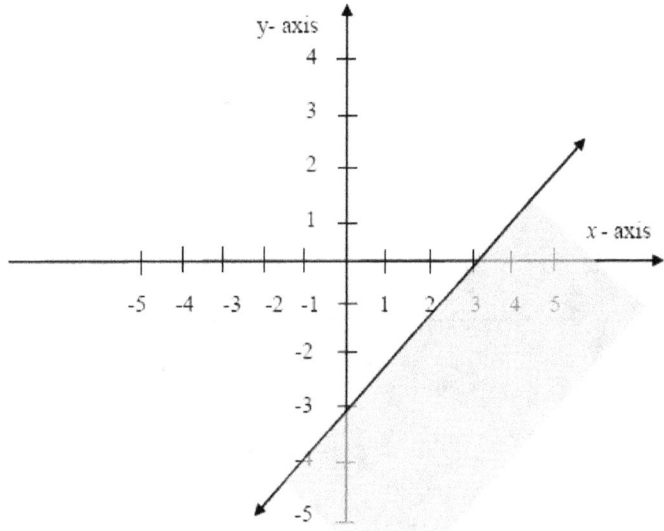

Problem No. 16

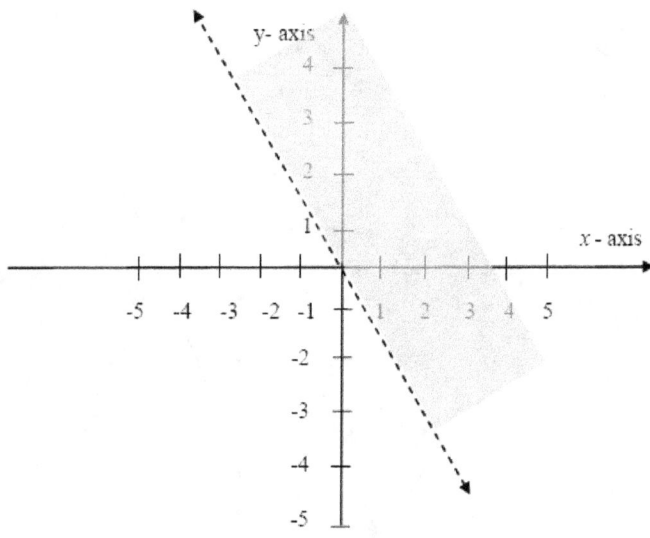

Exercise 11.4

Problem No. 17

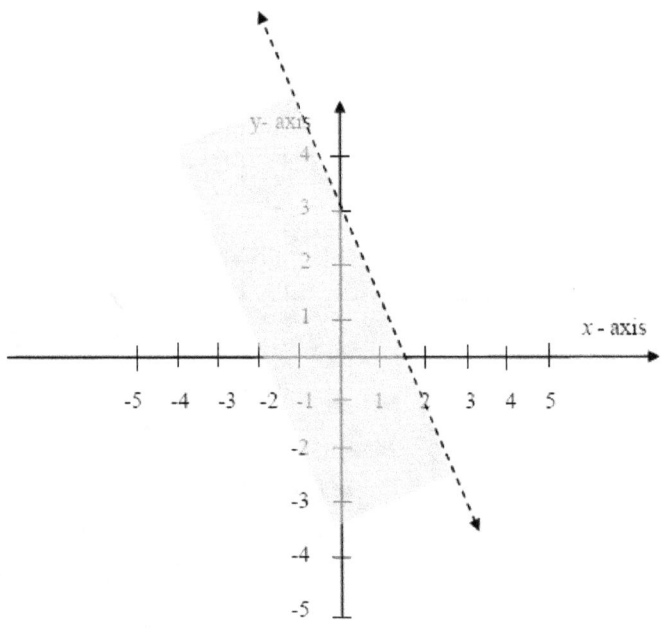

Problem No. 18

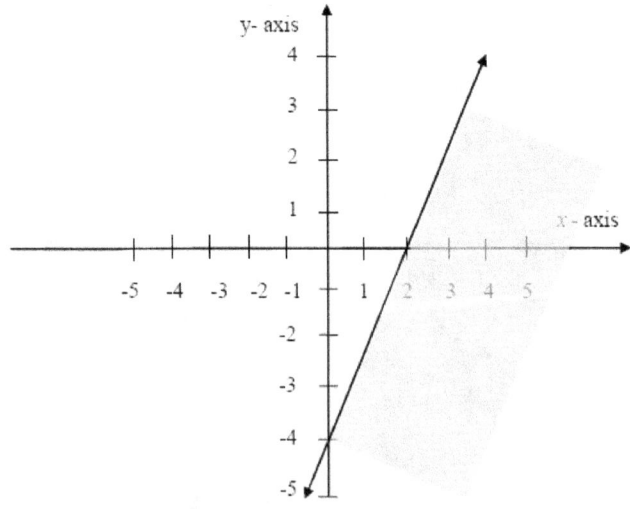

Exercise 11.4

Problem No. 19

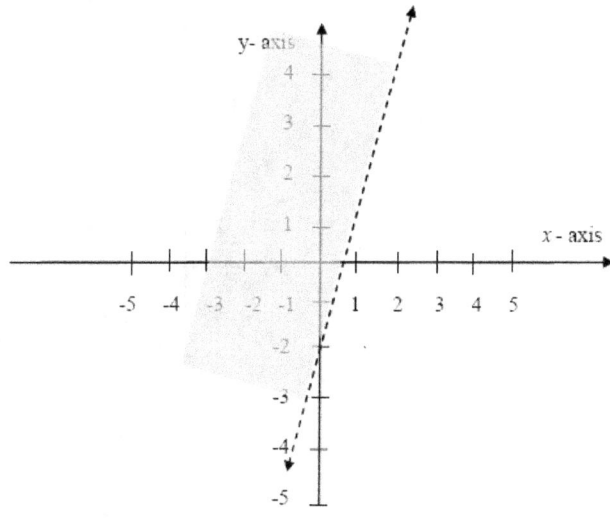

Problem No. 20

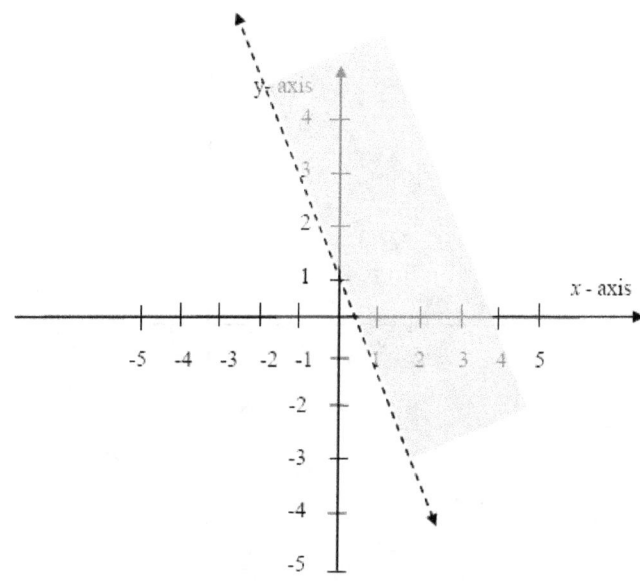

Exercise 11.4

Problem No. 21

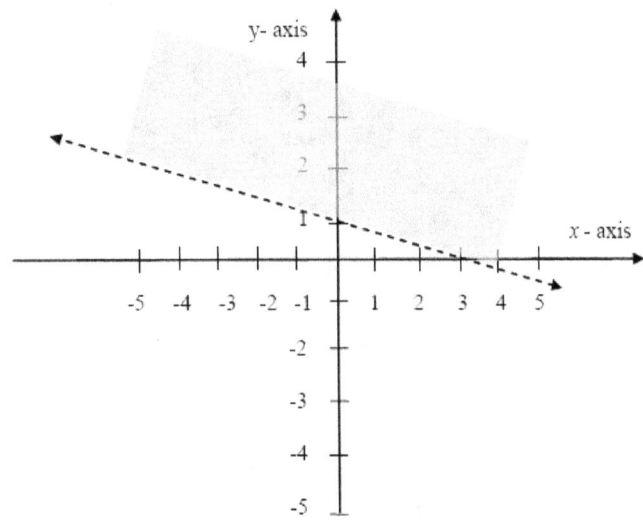

Problem No. 22

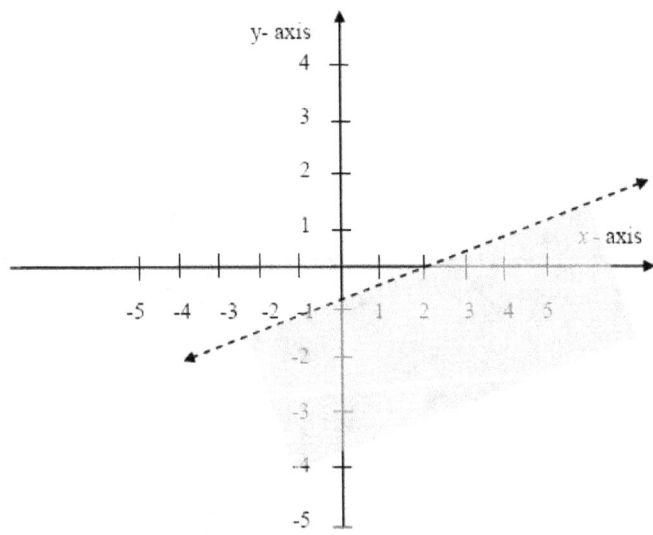

Exercise 11.4

Problem No. 23

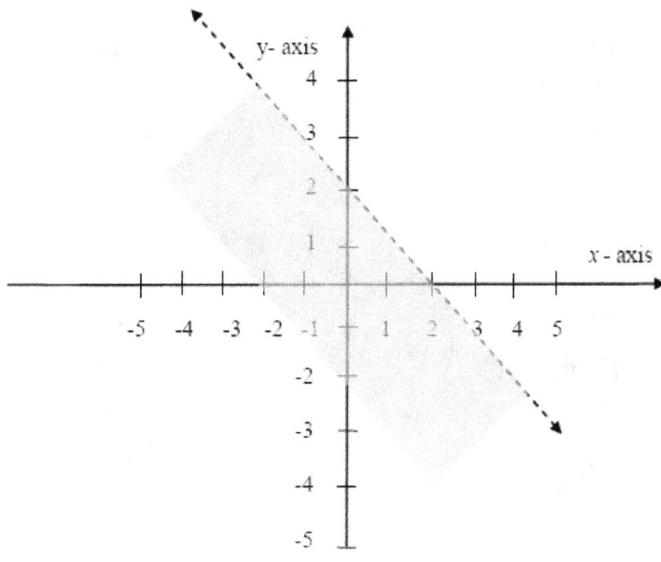

Problem No. 24

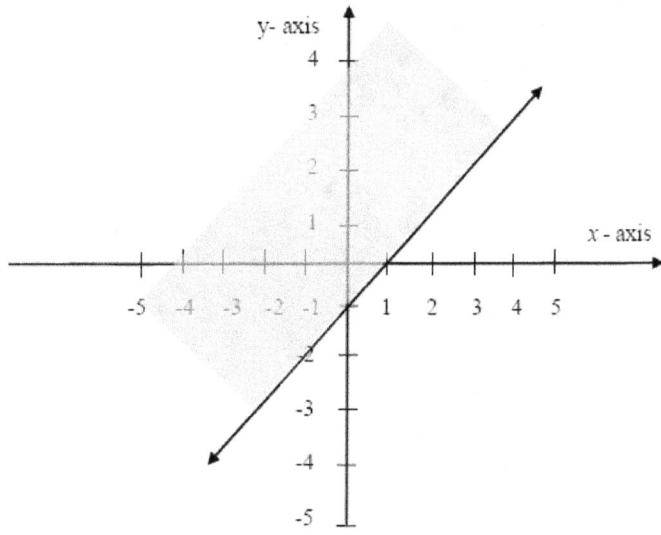

Exercise 12.1

1-	Top	2-	Top
3-	Bottom	4-	Bottom
5-	Top	6-	Bottom
7-	Top	8-	Bottom
9-	Bottom	10-	Top

Exercise 12.2

1-	Right	2-	Left
3-	Right	4-	Left
5-	Right	6-	Left
7-	Right	8-	Left
9-	Right	10-	Left

Exercise 12.3

Problem No.	Coordinates of vertex	Equation of axis of symmetry
1	$(0, 0)$	$x = 0$
2	$(0, 0)$	$x = 0$
3	$(0, 4)$	$x = 0$
4	$(0, 3)$	$x = 0$
5	$(2, 0)$	$x = 2$
6	$(1, 0)$	$x = 1$
7	$(3, -4)$	$x = 3$
8	$(-1/2, -1/4)$	$x = -1/2$
9	$(-2, 4)$	$x = -2$
10	$(-1, -2)$	$x = -1$
11	$(0, 0)$	$y = 0$
12	$(0, 0)$	$y = 0$
13	$(-5, 0)$	$y = 0$
14	$(12, 0)$	$y = 0$
15	$(5, -1)$	$y = -1$
16	$(-8, 1)$	$y = 1$
17	$(13/2, -3/2)$	$y = -3/2$
18	$(-7, 1)$	$y = 1$
19	$(-1, 3/2)$	$y = 3/2$
20	$(0, -1)$	$y = -1$

Exercise 12.4

1- $(1, 0), (-1, 0)$ 2- $(3, 0), (-3, 0)$
3-   The parabola does not intersect with the x - axis
4-   The parabola does not intersect with the x - axis
5- $(0, 0), (3, 0)$ 6- $(0, 0), (-4, 0)$
7- $(2, 0), (1, 0)$ 8- $(2, 0), (-1, 0)$
9- $(4, 0), (2, 0)$ 10- $(5, 0), (-3, 0)$

Exercise 12.5

1-	$(0, 2), (0, -2)$	2-	$(0, 4), (0, -4)$
3-	The parabola does not intersect with the y- axis		
4-	The parabola does not intersect with the y- axis		
5-	$(0, 2), (0, -2)$	6-	$(0, 5), (0, -5)$
7-	$(0, 2), (0, 1)$	8-	$(0, 3), (0, -1)$
9-	$(0, -4), (0, 2)$	10-	$(0, -5), (0, -3)$

Exercise 12.6

Problem No. 1

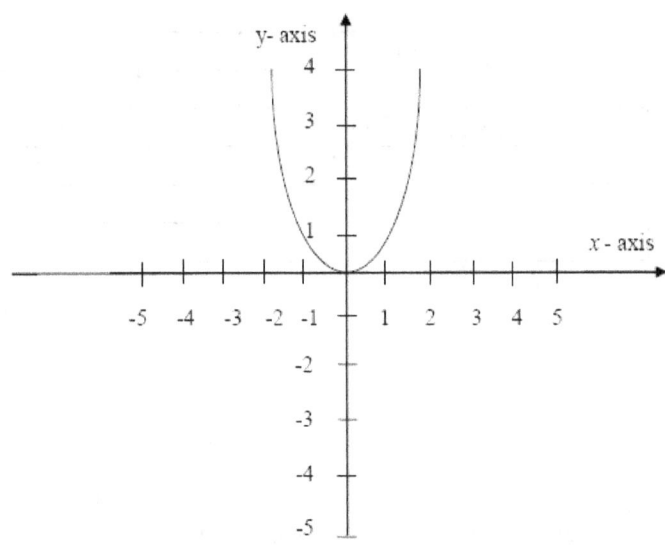

Exercise 12.6

Problem No. 2

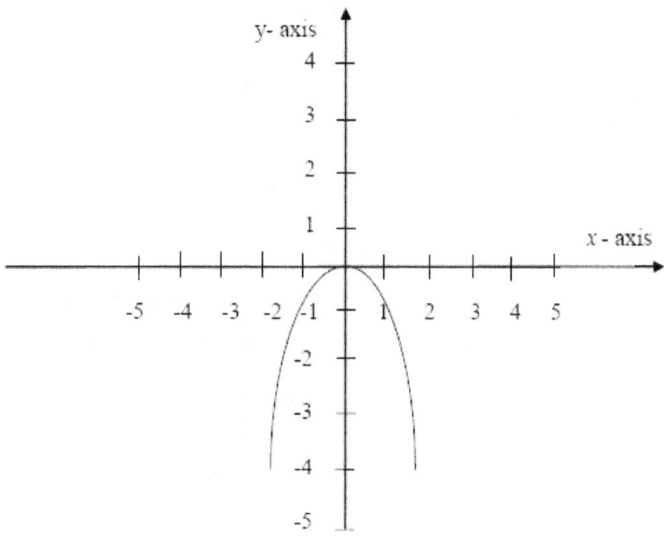

Problem No. 3

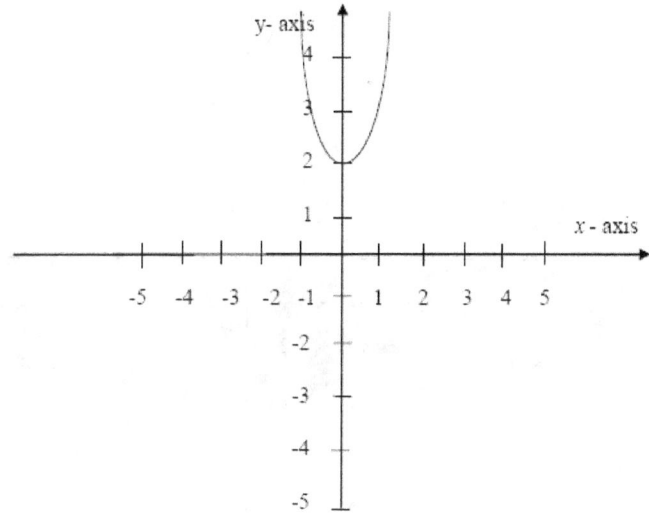

Exercise 12.6

Problem No. 4

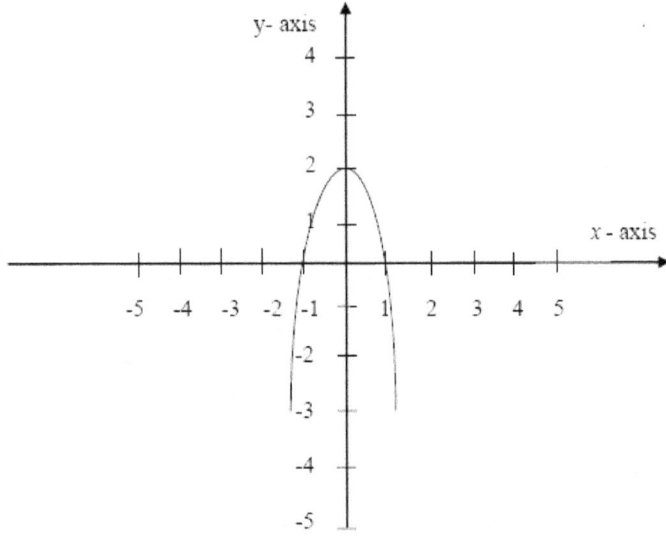

Problem No. 5

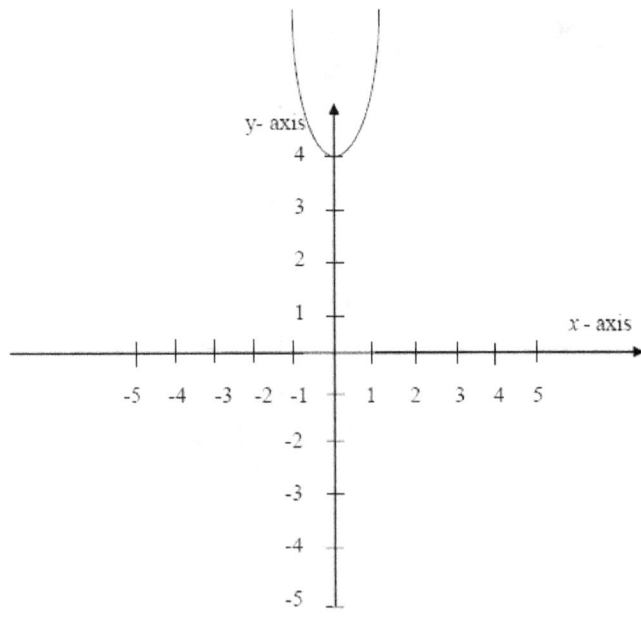

Exercise 12.6

Problem No. 6

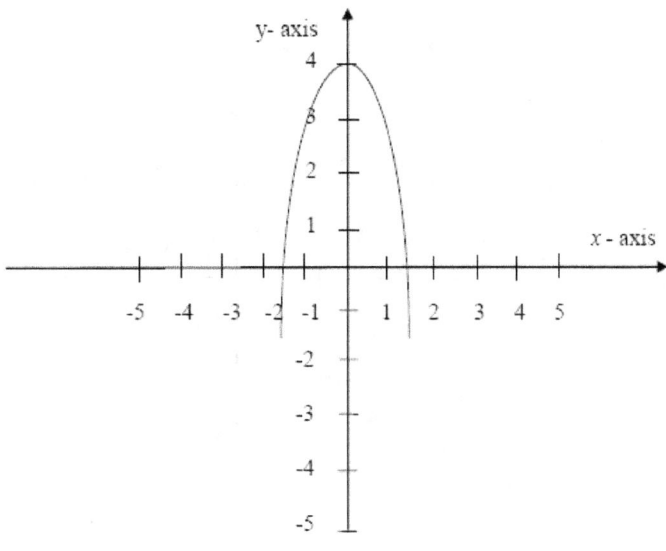

Problem No. 7

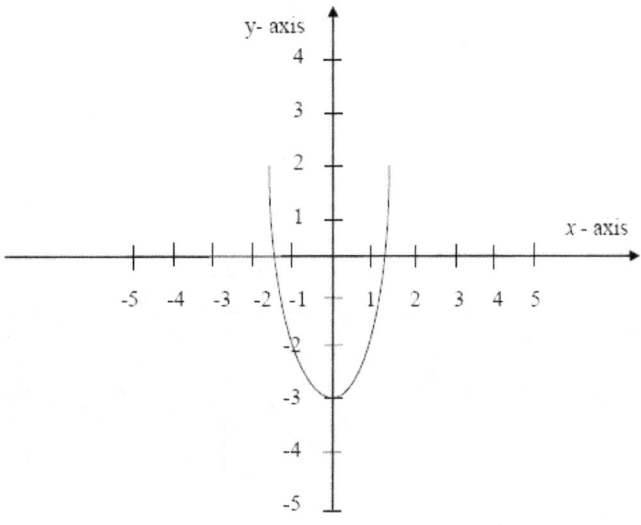

Exercise 12.6

Problem No. 8

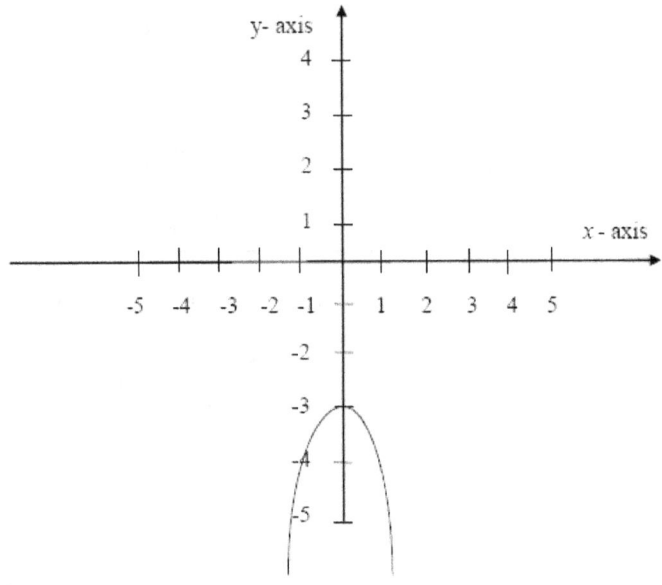

Problem No. 9

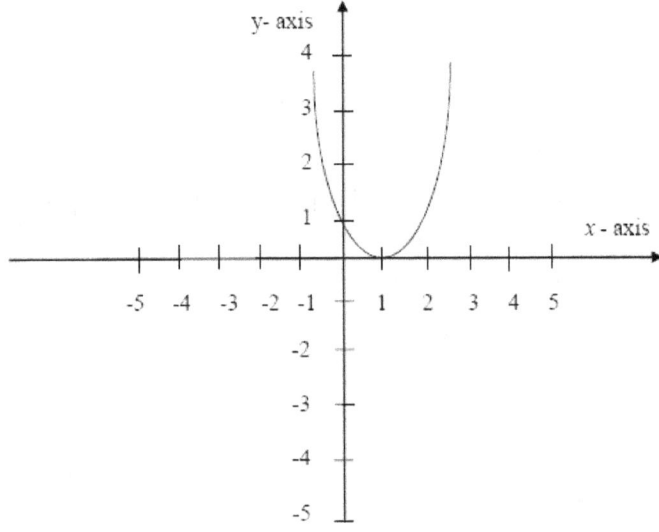

Exercise 12.6

Problem No. 10

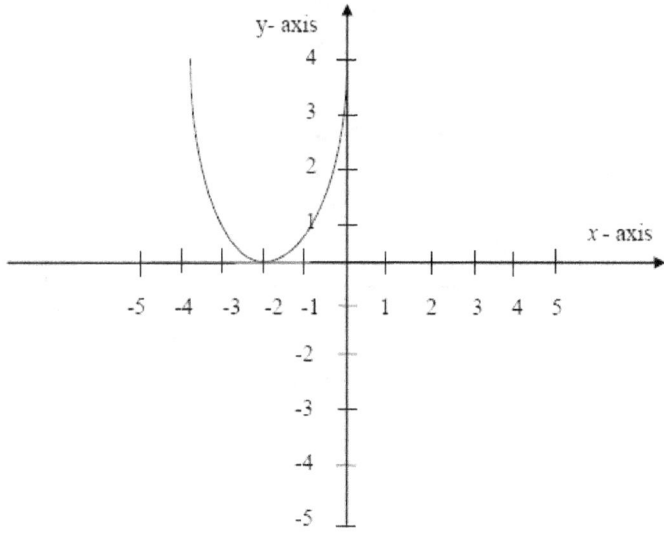

Problem No. 11

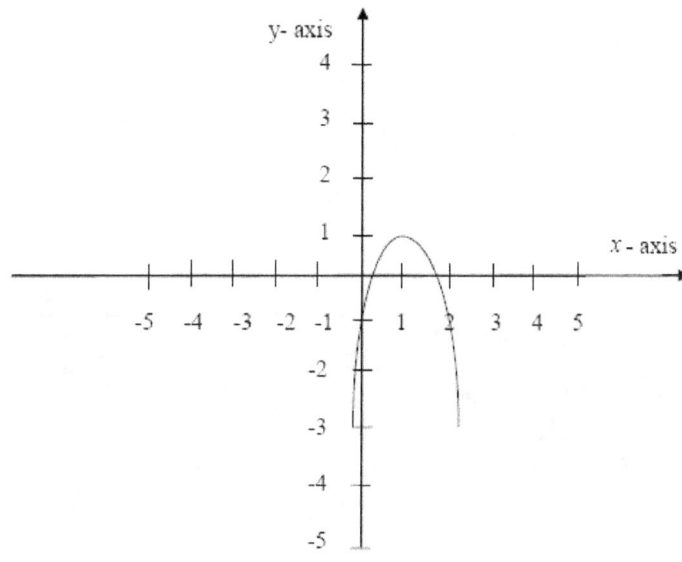

Exercise 12.6

Problem No. 12

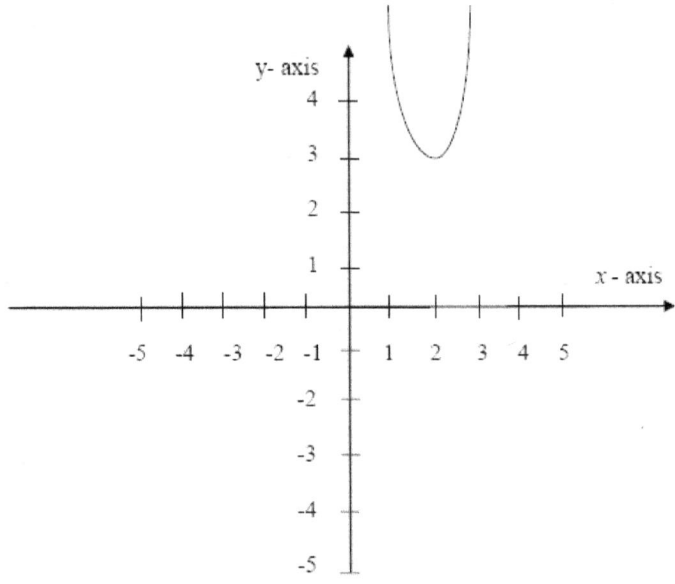

Problem No. 13

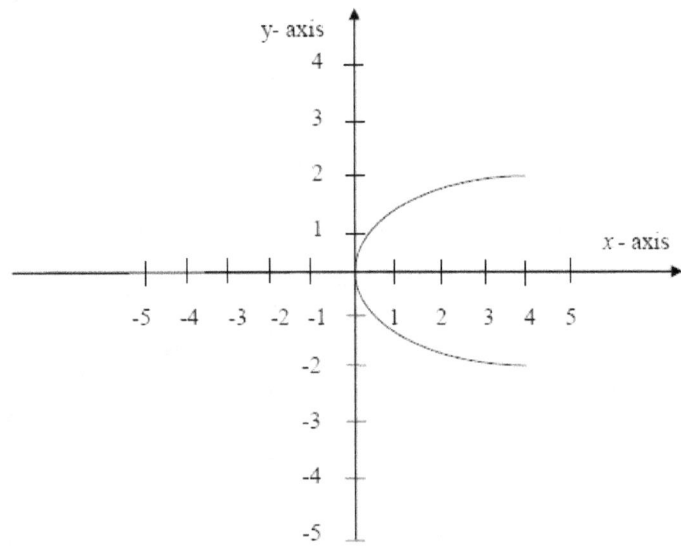

Exercise 12.6

Problem No. 14

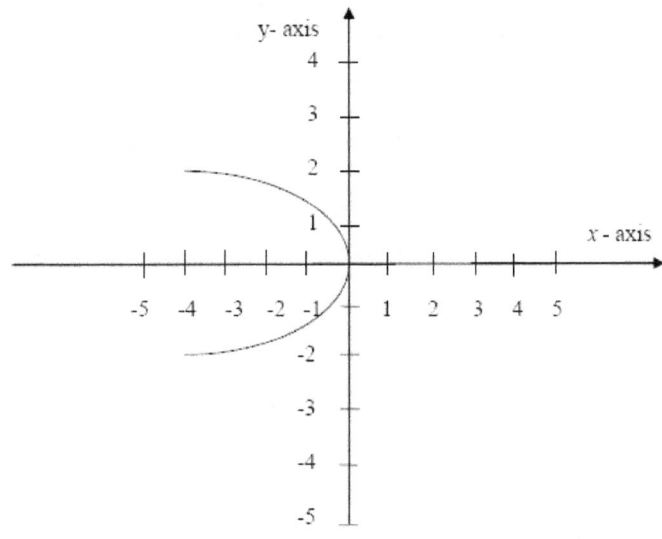

Problem No. 15

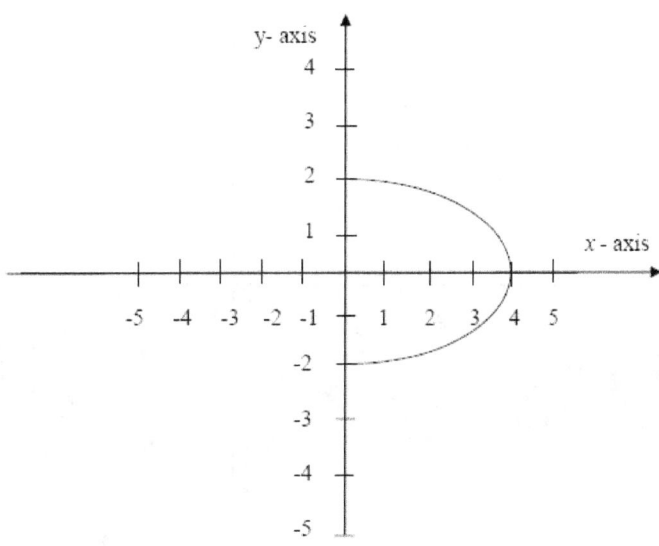

Exercise 12.6

Problem No. 16

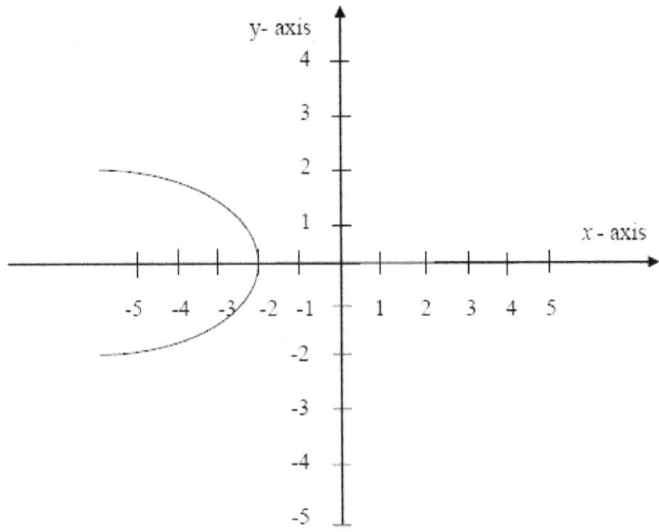

Problem No. 17

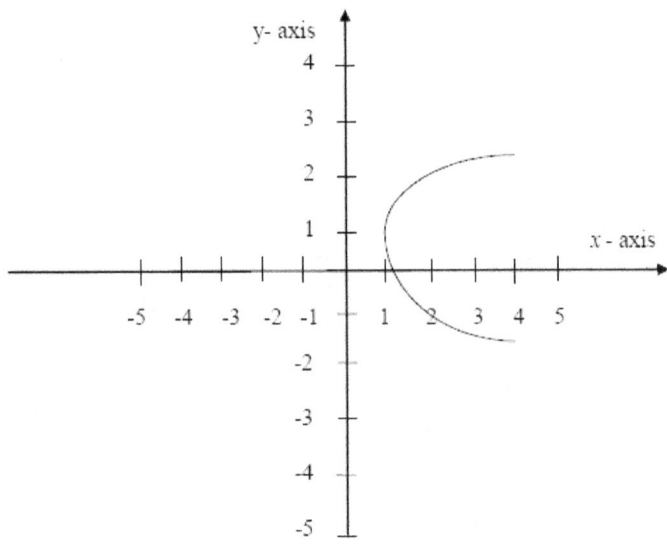

Exercise 12.6

Problem No. 18

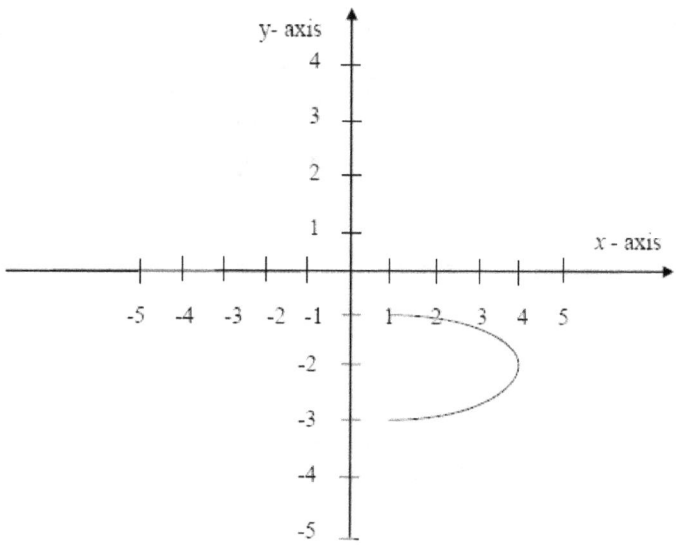

Problem No. 19

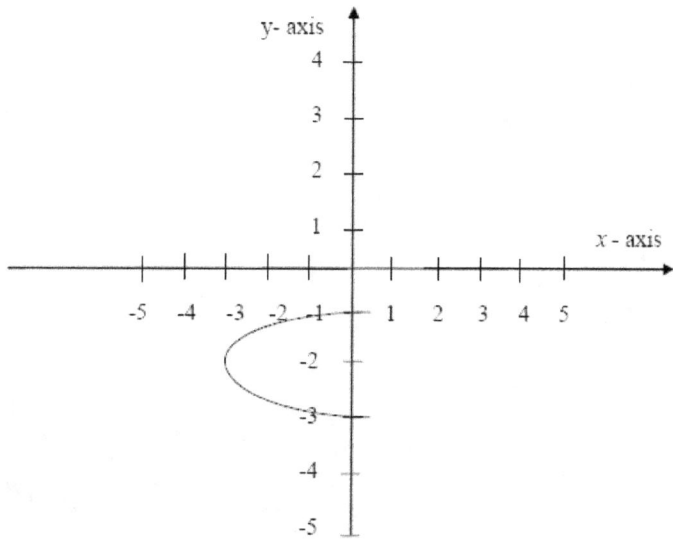

Exercise 12.6

Problem No. 20

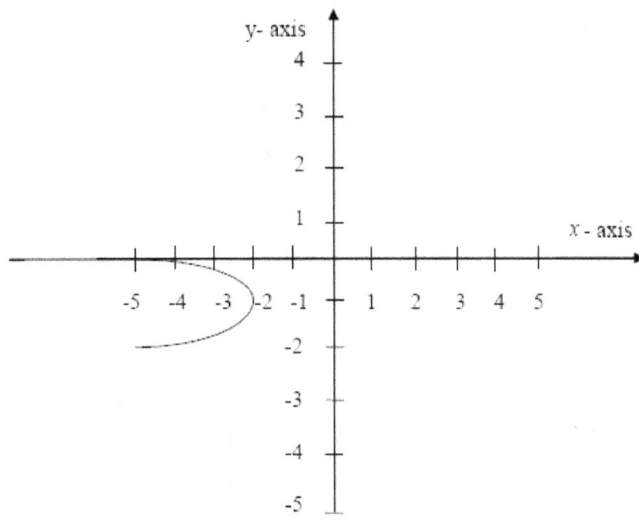

INDEX

www.ingramcontent.com/pod-product-compliance
Lightning Source LLC
Chambersburg PA
CBHW081108170526
45165CB00008B/2369